T0188018

DOMESTICATION

**Recent Titles in
Greenwood Guides to the Animal World**

Flightless Birds
Clive Roots

Nocturnal Animals
Clive Roots

Hibernation
Clive Roots

Animal Parents
Clive Roots

Domestication
Clive Roots

DOMESTICATION

■ Clive Roots

Greenwood Guides to the Animal World

GREENWOOD PRESS
Westport, Connecticut · London

Library of Congress Cataloging-in-Publication Data

Roots, Clive, 1935–
Domestication / Clive Roots.
    p. cm. — (Greenwood guides to the animal world, ISSN 1559–5617)
    Includes bibliographical references and index.
    ISBN 978–0–313–33987–5 (alk. paper)
1. Domestication—History. I. Title.
SF41.R66 2007
636.009—dc22   2007016145

British Library Cataloguing in Publication Data is available.

Library of Congress Catalog Card Number: 2007016145
ISBN-13: 978–0–313–33987–5
ISBN-10: 0–313–33987–2
ISSN: 1559–5617

First published in 2007

Greenwood Press, 88 Post Road West, Westport, CT 06881
An imprint of Greenwood Publishing Group, Inc.
www.greenwood.com

Printed in the United States of America

The paper used in this book complies with the
Permanent Paper Standard issued by the National
Information Standards Organization (Z39.48–1984).

10   9   8   7   6   5   4   3   2   1

For Jean
For the many years of love, companionship,
and shared concern for animals,
from London Zoo to Vancouver Island

Contents

Preface

Evolution is a process of natural change resulting from competition and survival. Since life began on earth, escaping predators and competing for food, shelter, and mates required continual improvements to an animal's lifestyle, and traits that aided survival were passed on to its offspring. The genetic composition of a population slowly altered as each generation improved to cope with new conditions, and eventually these changes affected their appearance, internal functions, and behavior. In time these natural changes produced new species, some so superbly designed, and their adaptations so successful, that their evolutionary development stopped and they have remained the same for millions of years. But they are the exceptions, as most could not compete, and extinction from natural causes became an integral part of evolution.

Although rapidly reproducing microscopic organisms can mutate quickly to challenge a new drug, change in the animal kingdom is constant but very slow. Major changes—for example, from amphibians to reptiles, and from dinosaurs to birds—took millions of years. The first single-celled life forms originated between three and four billion years ago, but the land vertebrates began to evolve only about 400 million years ago. The process was so slow that no one could expect to see the development of major new animals in their lifetime.

But times have changed. In the last century, while natural evolution continued its slow and unnoticed course, another form of evolution was happening simultaneously. It was totally unnatural and extremely rapid, and it has already produced new animals, many that differ considerably from their ancestors. The process is actually a very familiar one, at least in its original, prehistoric form, for it began when cavemen tamed the wolf. It is the controlled breeding of wild animals by man, known as domestication.

Introduction

Until recently evolution was solely a natural process, where only the fittest survived. The losses were very high, for the fossil record shows that many animals could not compete, and the current living species may represent only 1 percent of all the vertebrates that have ever existed. This natural improvement of wild animals, resulting from evolutionary chance, continues at its very slow pace, but for several millennia a parallel process unconnected to survival has also changed animals. Called domestication, after *domesticus,* Latin for home, it resulted from man's continual control of wild animals and their selection for certain characteristics. Domesticated animals are therefore man's creations, and the process has produced many new animals, adapted to artificial conditions, and reliant on man for their food, shelter, and mates. The controlled animals eventually changed genetically, and this produced detectable differences between them and their wild ancestors.

Although many animals had the opportunity to become domesticated, until the twentieth century man could not overcome the behavioral and physiological attributes that prevented the regular breeding, or even the maintenance, of some species. The original domesticates bred easily and were mainly social creatures, accustomed to living in groups and therefore more amenable to human control and association. Their gregarious nature allowed them to establish positions within the hierarchy of the group rather than needing to defend territories of their own. But although a social (herd or pack) structure favored domestication, it was by no means a prerequisite, as so often stated, for the African wild cat (ancestor of the cat), and the polecat (ancestor of the ferret) are certainly not social animals. Even golden hamsters are actually bad-tempered, unsociable creatures, but they are more recent domesticates. Consequently, until the last century domesticated animals were restricted to our familiar pet and farm animals, such as dogs, cats, cattle, sheep, horses, chickens, and rabbits, plus a few lesser-known creatures like the ferret and guineafowl. These animals have been domesticated for millennia. The wolf was the first to be controlled, presumably by cavemen, and others followed as man settled down and began cultivating crops and herding animals like sheep and goats. Great changes resulted from this long control, and many breeds of domesticated animals differ considerably from their ancestral stock, in size and shape (morphologically), in the manner in which they function (physiologically), and certainly in their behavior.

The desirable characteristics for domestication were an animal's value to man for supplying food, skins, and fiber, for draft, and for hunting. They had to be easy to feed, and those able to forage for themselves—herbivores such as cattle, sheep, and guinea pigs; and omnivores like the duck, junglefowl and wild boar—were prime candidates for control. A calm and tractable disposition was necessary, as was a temperament that allowed them to live in close proximity to man and to breed readily in an artificial environment. But there were obligations on the keeper's side also, for domesticated animals are completely dependent on humans, who therefore have a moral duty to provide the best possible care.

Surprisingly, between the control of the easily bred animals all those years ago and the beginning of the twentieth century, a period of several millennia, the domestication of new animals did not continue at the same pace; and there is a great gap between the old domesticates and the new ones. Attempts were made in ancient Egypt to domesticate many animals, including antelope, cranes, ibis, baboons, and perhaps even hyenas, although it is difficult to understand why, but none were successful. The turkey, canary, peafowl and ring-necked pheasant were the major species controlled during that period, when emphasis was placed upon improving the existing domesticated animals to better suit man's purposes, rather than creating new ones.

■ RECENT DOMESTICATION

After centuries of inaction in the domestication of new species, during the last century, and especially in its latter half, there was an unprecedented escalation of controlled animal breeding that produced many changes. Dozens of animals that had rarely bred before began to reproduce regularly. It became relatively easy to breed pet and hobby species such as tree frogs, boa constrictors, bearded lizards, scarlet macaws, fennec foxes, and even hedgehogs. On farms, red deer, mink, and catfish were selectively bred—intentionally mated to achieve or eliminate a specific trait. Many wild animals, including shrews, geckos, and even pit vipers, were maintained for numerous generations in research facilities. The new age of domestication was well under way, in two basic forms. There was the rapid commercial process intent on producing change, and the zoo process, just as intent on keeping animals the same. Animals considered difficult to breed at the beginning of the century were producing new mutants of color and pattern by its end, most of them more valuable than their natural ancestors. Many zoo mammals began reproducing regularly, including even gorillas and rhinoceroses, and the breeding of some species had to be curtailed due to the lack of space. These animals still resemble their wild ancestors externally in keeping with the aim's of the zoo profession.

There were several reasons for the accelerated reproduction and domestication of wild animals in the twentieth century. Many new species were available and accessible, and shipping by air simplified their acquisition. Prior to the introduction of conservation and animal health legislation, animals could be freely exported from many countries, and species traditionally considered zoo animals were available to private breeders. Improved nutrition and husbandry resulted in increased breeding—the first and most essential step in the domestication process. This recent phase of wild animal control also benefited from a number of artificial breeding practices, especially planned inbreeding, linebreeding, hybridizing, artificial insemination, and the deliberate production and perpetuation of mutants.

Whereas certain criteria applied to the early domestication of wild animals, such as gregarious behavior, nonaggressiveness, and acceptance of replacement diets, those factors do not apply to the new domesticates, where the only applicable consideration is a willingness to breed in man's care. An animal's social behavior is irrelevant. Solitary animals like the cheetah, snow leopard, mink and Arctic fox are all becoming domesticated. An animal's diet no longer impedes its control by man, and commercial diets are now available for many species including hedgehogs and caracals; while mink and foxes are even fed a paste of blended offal from food processors and slaughter houses. An animal's aggressive nature is no longer a deterrent, and several species quite capable of killing humans are now being domesticated, including elk and bison, crocodiles, giant constrictors and pit vipers. One aspect that still applies, at least to the speed with which domestication can proceed, is rapid growth and early reproduction. But elephants and giant tortoises have also begun the domestication process, and their slow growth rate will be more than compensated for by their long breeding life. The process will simply take longer.

Breeding is the first stage in the process of domestication, which otherwise cannot proceed. When captive animals breed continually, domestication is unavoidable, whether in the zoo, deer farm, fur farm, pet-breeding establishment, or research laboratory. Two main factors aid the process. First, the change in an animal's environment from wild and free to one of confinement, including the disruption of its social and sexual life, and the change of food. Second, the practice of artificial or selective breeding—the intentional selection of traits and strains for a specific purpose—to change the characteristics of the captive animals. The general rule of domestication is that existing characters are selected rather than new ones being produced. Artificial selection exploits a wild animal's genes, producing the basis for change from their wild ancestors, and eventually resulting in their dependence on man and often their unsuitability for return to the wild.

The escalation in the breeding of wild animals was powered by two differing motives. Commerce was the driving force behind the great increase in new pet and hobby animals, and many are now produced in numerous mutants. In the mid-twentieth century, captive parrots rarely bred, mainly because they could not be sexed and pairing was a matter of guesswork. They are now easily sexed by DNA testing of feathers or blood, and hundreds of breeders now produce thousands of hand-raised pets annually. Falcons were also very difficult to breed, but given the isolation most of them need they now reproduce readily, and are even hybridized in an attempt to produce better birds for falconry. Commercial reptile breeders produce many color and pattern mutants of corn snakes, ball pythons and leopard geckos.

Unfortunately, certain risks are associated with domestication, and inbreeding is a major one. It results in lowered fertility and virility, poor viability of the young, and the perpetuation of unnatural characteristics, which may be undesirable. However, desirability depends on the breeder. While many commercial breeders prefer to produce mutants, the natural phenotype of the wild animal is most desirable for zoos and some private breeders. Color or pattern mutants (which are phenotypic aberrations) are normally not perpetuated, except for such prized zoo exhibits

Hybrid Peregrine Falcon and Saker Falcon *Considered difficult to breed until just a few decades ago, the decline of wild falcons due to organochlorine pesticides resulted in intense institutional efforts to breed birds for reintroduction. Private breeders then began producing birds for falconry, and now hybridize several species in an attempt to improve on nature.*
Photo: Clive Roots

as white tigers and white lions, and the king cheetah, a pattern mutant, now being bred at the De Wildt Cheetah Centre in South Africa.

Zoos and related institutions, such as bird gardens, wildlife parks, and reptile parks, were also involved in this wave of breeding success, but their sights were focused mainly on endangered species conservation and on reducing their former reliance on wild animals to stock their enclosures. They were so successful in breeding some species, and consequently flooding the market for acceptable homes, that breeding was eventually curtailed. A comparison of breeding results for the years 1950 and 2000 clearly show how animal reproduction increased in both species and numbers during this period. Research establishments also contributed to the escalation of captive breeding and its resultant domestication, maintaining many animals such as newts, shrews, hedgehogs, and mouse lemurs in long-term breeding colonies. In all these animals, however, the quality and characteristics of the offspring were influenced by artificial selection, both of the composition of the breeding pair or group and of the desirable traits to be perpetuated. In small captive populations, this can result in the establishment of pathological characters, whereas the normally large numbers of animals in the wild prevent this.

Commercial breeders often attempt to produce animals that differ in some way from their wild ancestors, whereas noncommercial breeders try to maintain animals as close as possible in their characteristics to their wild ancestors. However, just as there are selective forces at work in the wild, so there are pressures on all captive animals whoever breeds them, including unnatural housing, unsuitable nutrition, inability to engage in their normal social behavior, and artificial mate selection; and these forces similarly result in change. Also, selection always occurs, even if it is neither intended nor preferred. Poor breeders are automatically culled in favor of animals that thrive and reproduce regularly, thus perpetuating those that more readily accept human control and artificial conditions, and continuing the domestication process. In the commercial sector selection may then become intentional, with animals that have desirable characteristics or behavior being selected for breeding, and further involvement in the process may include deliberate inbreeding, linebreeding, hybridizing, and the production of mutants. Zoos differ in their selection protocol. Some of the world's best zoos are producing phenotypic mutants, such as black jaguars, white lions, and white tigers—resulting from continual inbreeding—purely because they are crowd-pleasers and thus revenue earners. But most zoos, with the involvement of species survival plans, studbooks, and the help of geneticists and population managers, aim to preserve the genetic integrity and purity of their stock, but they cannot avoid its domestication.

Taming an individual animal is not domestication. Hannibal's African elephants and the Asiatic elephants that have been used for centuries for war, work, and ceremony, were tamed, not domesticated, as they were caught in the wild as calves, raised and trained. Only in recent years has the domestication process begun as elephants are beginning to breed more frequently in camps, rescue and rehabilitation centers, and zoos. The value of tameness depends entirely on the purpose of the animals. Taming is not mandatory for the Siberian tiger or the polar bear, whose zoo enclosures will not be entered when they are in residence. The farmed red fox should accept the closeness of humans, but need not be hand-tame like the fennec fox cuddled in the home. Zebras have occasionally been tamed and trained to pull carts, purely as a novelty, even in "reputable" zoos years ago before such activities were frowned upon; but attempting their full-scale domestication makes no sense when horses, asses and mules adequately supply man's equine needs. Therefore, until now, zebras have never been domesticated, at least to act as horses, but the process is now well underway in zoos, where Burchell's and Grevy's zebras have been kept and bred for many generations. But they can never act as their wild and free ancestors.

Changes to an animal through taming are primarily behavioral, in particular the reduction of its flight reflexes. Although taming to the point of docility and ease of handling is not essential for domestication, it certainly helps the process, for nervous, aggressive carnivores and "flighty" antelope and deer are less likely to breed. Whereas a degree of tameness is preferred for farmed red deer, complete tameness is mandatory for the pet macaw or cockatoo, which is best achieved through hand-raising. The natural nervousness of wild-caught mammals is rarely seen in zoos these days anyway, as most are captive-born. Animals whose flight reflexes have been reduced or eliminated, and that accept control and the closeness of man without recourse to attack or flight, are obviously better candidates for domestication.

The process of domestication varies in speed according to the species and the extent of its control and reproduction, but it is certainly aided by a short breeding cycle. The female golden hamster, for example, is sexually mature when 10 weeks old, and has a gestation period of just 16 days, which can result in three generations in one year. More generations annually means quicker domestication, and quicker change. It is also aided by the small size of the founder population, such as the single family of golden hamsters captured in Syria, and the consequences of inbreeding and genetic drift. Inbreeding is encouraged by commerce, for the mutants that sooner or later appear are usually valuable and are perpetuated, leading to the production of strains with new colors and patterns. The race to mutate has new mutants of some species, such as corn snakes and leopard geckos, appearing almost monthly.

In addition to all the new domesticates, the last century saw the commencement of the redomestication of many of the ancestors of our familiar breeds, but this time in their natural, pure form. Wild boar, whose control about 7000 BC resulted in the numerous breeds of domestic pigs, are now farmed for their meat. The mouflon and wild goat, ancestors respectively of the sheep and goat, have been kept in zoos and wildlife parks for many generations as pure, characteristic species. Similarly, the ancestors of many other long-domesticated animals, including the red junglefowl (chicken), rock pigeon (pigeon), mallard (duck), graylag goose (goose), and even the wolf, are all being kept and bred in their natural wild form. For many animals there are now two distinct domesticated populations—the original prehistoric one, in all its varying breeds, and the recent unchanged ones.

In contrast, it seems unlikely that some animals will ever be domesticated. Species like the tamandua anteater, the yapock or water opossum, and the insectivorous bats—whose husbandry has yet to be mastered—surely cannot be domesticated. They have no commercial value and seem unlikely to be good pet subjects. But similar improbable animals—shrews for example, and even newts—have been maintained in laboratories for many generations. Who would have thought, just 50 years ago, that hedgehogs, fennec foxes, bearded dragons, and ball pythons would be domesticated for the pet trade; that zoos would be breeding gorillas and rhinos regularly, or that farmer's fields would be stocked with ostriches and hybrid deer instead of cattle and sheep?

■ THE DOMESTICATORS

Three factions contribute to the current domestication of wild animals. They are the commercial breeders, zoos and private breeders, and research establishments, and their aims, and therefore their methods, differ widely. The commercial breeders include the farmers of wildlife, who produce animals for their products. They include fur farmers producing new shades of mink; game farmers who cross red deer and elk to produce animals with larger antlers, and ostrich farmers who select their stock for improved egg production and bigger thighs. The business of these animal breeders is therefore largely terminal. The commercial producers also include the pet trade suppliers of parakeets, parrots, lovebirds, finches, snakes and lizards, now available in many new color and pattern mutants. Others supply "home-bred" monkeys and hedgehogs, and many breeders produce new mutants of pheasants and waterfowl, and hybridize falcons. Unlike man's early association with animals—which is assumed to have happened accidentally, and then eventually led to an appreciation of their value and thus even greater control—the value of these new animals is readily appreciated and modern commercial domestication is a growth industry involving their large-scale production. As in the wild, where natural selective forces encourage change in animals, the artificial forces that exist in the captive environment also produce change.

Whereas natural evolution was for the animal's benefit, its improvements associated with survival, commercially domesticated animals are changed to serve a specific purpose, always for man's benefit. It therefore follows that any changes that may increase the benefits will be vigorously pursued. Commercial breeders are striving for agricultural and pet objectives—production efficiency, rapid growth, breed improvement, and new breed development—for economic reasons. They select their breeding stock for size, growth rates, and egg and meat production, and they seek new mutants and personality traits such as docility and acceptance of human companionship or control. Artificial selection and inbreeding are now practiced by the breeders of pythons, parrots, and foxes, just as the breeders of Rhode Island chickens, Merino sheep, and Irish wolfhounds have done for centuries. Most new forms of domesticates are still known by their color or pattern, such as albino python or blue ring-necked parakeet, and have yet to be given "breed" names such as those of the old domesticates, like bulldog and borzoi. An exception is the teddy bear hamster.

In contrast, zoo breeding is now far removed from commerce, for the intention is to keep species pure and characteristic of their wild ancestors. Hybridization and, in most species, the perpetuation of mutants, are avoided, and in-breeding due to small populations is carefully monitored. Studbooks and cooperative breeding programs assist in these endeavors, their goal being the long-term preservation of the natural species, especially in view of the possibility of their eventual return to the wild. However, change occurs through the selective pressures placed on animals by their environment, although there may be no deliberate attempt to change them. Although they are being kept pure and still resemble their ancestors, zoo animals cannot avoid becoming domesticated.

■ THE CHANGES

When animals are continually controlled by man, most aspects of their life are artificial. Many can no longer select their own mates, guard a territory, or battle with conspecifics for the control of the herd. Their movements are restricted; they

Barbary Sheep or Aoudads   *Aoudads are not true sheep, but an intermediate species between the sheep and the goats, which have shown an amazing ability to withstand the effects of intense inbreeding. All the animals in British zoo collections are descended from a few imported by the London Zoo in 1842. They are still generally characteristic of the species, even though they are now considered homozygous or genetically identical.*
Photo: Carolyne Pehora, Shutterstock.com

cannot migrate and usually cannot hibernate. They cannot hunt, and their food selection is limited, often unvarying, and may not be suitable for their teeth and digestive systems. They become domesticated and change occurs, compounded by contrived mating arrangements, especially artificial selection, inbreeding, and hybridization. The changes that occurred in the original domesticates millennia ago are now being repeated in the recent ones. Originally, long coats were desirable in certain breeds of sheep for commercial reasons, and in cats for visual appeal; now, long-haired hamsters are favored over their short-coated ancestors, and angora ferrets are being produced. The many breeds of white cattle, goats, dogs, and chickens produced long ago are now mirrored by more exotic animals such as white pythons, geckos, wallabies, and lions.

Domestication results in changes to an animal's phenotype—its observable characteristics—and these changes may be behaviorally or genetically induced. Morphological phenotypes relate to the animal's form (its physical shape and size), while the behavioral phenotype is manifested by its actions or reactions in response to external or internal stimuli. After generations of regular selection for behavioral phenotypes, and most likely for morphological ones also, animals differ from their ancestors, and their genotype (their genetic makeup) will also differ. By favoring suitable traits, especially ones that will improve the breeding record, and by providing extra care to save weaklings, breeders may raise animals with deleterious genes that would have caused their demise in nature. In zoos, where a major justification is the eventual return of animals to the wild, these changes will unfortunately render them unsuitable for reintroduction into their former habitat without considerable "remodification," which may be impossible anyway as it would likely require an infusion of genes from their wild counterparts.

None of the new domesticates have been subjected to the long-term control of the familiar pet and farm animals, and obviously in such a short period—just half a century for many of them—they have not experienced the same degree of change. With few exceptions, such as the budgerigar in which British show birds are considerably larger than their wild ancestors, the new domesticates differ externally from their predecessors only in color, coat or pattern. They do not yet show the size variation of the longer-domesticated dogs, cattle, and horses. Consequently they are still known by names that reflect these initial changes, such as albino or blue, striped or stripeless, rather than old breed names that mostly reflect their origins, such as Yorkshire terrier, Manx cat, and Leghorn chicken. However, domestication produces such dramatic changes to an animal's morphology and physiology that most of the long-domesticated species are considered new species or subspecies. For example, the domesticated dog (*Canis familiaris*) is considered sufficiently different from the wolf (*Canis lupus*) to warrant specific status, but the chicken (*Gallus gallus domesticus*) is a subspecies of the red junglefowl (*Gallus gallus*). But these domesticates are then further subdivided into breeds—animals with uniform, inheritable characteristics—that are not geographically restricted. Breeds are therefore the domestic equivalent of wild animal subspecies—which are natural geographic divisions influenced by their environment. Like subspecies, breeds can interbreed.

■ THE ADVANTAGES OF DOMESTICATION

It has been suggested that the development of new domesticates has benefits for mankind, but these vary depending on the animals involved, and the greatest benefits will be from the new food animals. The world's human population is about six billion and is expected to grow at the rate of 75 million annually. Many are undernourished and the annual death rate from starvation continues to climb. Some animal domestication, such as the village industries in West Africa and Central America, where giant cane rats and iguanas, respectively, are "farmed" for the cooking pot, will help a little. But commercially produced specialty products such as venison, emu, and ostrich meat will neither feed the multitudes nor improve the plight of the world's poor. However, the development of new breeds and strains of the recently domesticated catfish, carp, tilapia, and other fish, resulting in greater and faster growth, will obviously reduce the strain on wild fish populations and significantly aid cattle, goats, sheep and pigs in supplying the world with essential protein. Where other animal products are concerned, the output of fur farms and crocodile farms may improve the quality of life for a few, but their distinctive mark on society is purely a cosmetic one. However, the escalation of captive breeding will certainly help some wildlife species. The recent boom in the desire for unusual or alternative pets, leading to the farming of iguanas as pets and the production of mutant snakes, lizards, and parrots, continues to reduce the once very large trade in wild-caught animals. Legislation introduced for conservation or animal and human health concerns has helped to reduce this once enormous trade in wild animals, but every alternate pet produced commercially further reduces the drain.

Unnatural evolution is escalating, and it is quite likely that natural evolution, as we have known it since Darwin's time, is doomed anyway. Every population of animals is affected by man's tremendous impact on the world. They have lost their habitat or have had it altered significantly. Whole populations have succumbed to climate change and to the effects of pollution and pesticides, overfishing, and overhunting. The selective harvesting of certain desirable species, such as Amazonia's spotted cats and the cockatoos of some Indonesian islands, seriously affects the natural balance, coupled with the loss of five million acres of rain forest annually. About three million square miles of the world's tropical rain forest remain, and at the current rate of deforestation they will be gone by the middle of the century.

Then there is the question of whether we can, or even need to, improve on nature. Our original domesticated animals show that it can be improved upon, at least for certain purposes of value to man. The food available as result of animal (and plant) domestication has been instrumental in the development of mankind and its cultures, and of course the massive population increases of recent centuries. Whether we need to improve upon nature even further or just concentrate on the existing domesticates is a far more complex question. Is a more colorful corn snake or a lutino ring-necked parakeet, an improvement? Some of the mutants now being bred are lovely animals, and are perfectly acceptable if their ancestors are allowed to continue life as normal in the wild, which is unfortunately becoming increasingly difficult for many species.

Zoos and similar wildlife breeders must continue to produce animals characteristic of their wild ancestors, even though they may otherwise be affected by man's control. Domesticated animals are acceptable replacements for wild animals, and for some species there is no alternative. Animals extinct in the wild, such as the European bison, Przewalski's horse, and Pere David's deer, have been totally controlled, domesticated and returned to suitable areas where they are now living semi-wild. Domestication is certainly not to blame for the current plight of the world's wild animals. In fact the world is a much richer place as a result of domestication, with hundreds of cattle, dog, sheep and chicken breeds, all derived in each case from just a single species. Domestication is therefore in some circumstances a counterpoint to extinction.

■ THE TIMING OF DOMESTICATION

The actual timing of human control of the original, prehistoric, domesticates has always been controversial, and in some cases may be out by a millennium, or even more. We have no absolute knowledge of the early development of the dog, and little of numerous other original domesticates; and every excavation of an ancient midden or burial site could conceivably change the dates. In contrast, there are quite accurate records of the first breeding of animals in the nineteenth and twentieth centuries, and therefore the beginning of their domestication. However, the controversy over the new domesticates will not be when the process actually began, but whether they can really be considered domesticated. There is no doubt that the golden hamster is domesticated, yet this familiar pet rodent was discovered only in 1930. The gerbil, another popular pet rodent, has only been kept since the 1950s in the laboratory, then later as a pet. Chinchilla farming began only in 1923 and they became popular pets much later. The farming of mink, Arctic fox, and red fox began only a century ago; although some may doubt their domestication, 100 years of selective breeding, and probably almost as many generations, has produced many mutants of more value to man than their original wild color. How quickly some animals can be domesticated was shown by Dmitri Belyaev, who kept silver foxes for 40 years and bred only the friendliest animals. After 18 generations, his foxes enjoyed human contact and showed many of the external characteristics of domestic dogs. Experiments with wild rats resulted in similar changes to their natural characteristics. First-generation captive-born brown rats were as territorial and aggressive as their wild caught parents, but after five generations—a period of just three years, as they are sexually mature six months from conception—strange male rats were welcomed by others in the typical domestic white rat manner.

There are, of course, degrees of domestication. Animals whose control by humans began several thousand years ago are obviously more domesticated than those that began to breed only in the last century. However, even for the original domesticates the process is not complete, and never will be. Domestication is a continual process. It starts when captive wild animals begin to reproduce regularly, and continues when future generations are just as productive. Whether it began several thousand years ago, or in the last century, it is a process of uninterrupted and unending improvement.

The first chapter of this book is devoted to the original or prehistoric domesticates, when and where they were first controlled by man, and their subsequent changes. The recent domesticates then follow, with chapters on each of the classes of vertebrates from fish to mammals. The zoo profession is discussed separately, as it is the only animal breeder involved neither in the escalation of unnatural evolution nor in the terminal use of animals. Finally, there are the domesticates that have regained their freedom, living off the land again in the manner of their wild ancestors.

1 Changing Nature

Until recently the time and place of animal domestication was estimated by studying their remains in prehistoric sites occupied by humans, determining the time of their death and the nature of their association with man. This method is now supplemented by studying the animal's DNA, which will help to determine its genealogy and its possible date of domestication. One of the difficulties in dating their remains, however, stems from the fact that the prehistoric domestication of wild animals is believed to have been accidental in most cases, at least initially, resulting from the social relationships of animals and man, rather than deliberate action to increase their food supplies, to carry them or their goods, or to act as companions. The people of the time had neither cages nor enclosures to control animals and no breeding stock they could pair off, so the process was obviously random and must therefore have been quite slow. Consequently, the timing of their domestication is quite vague and open to conjecture for most species, and the dates usually quoted may be off by many centuries.

The dog has always been considered the first of man's domesticated animals, and the process is believed to have started in the Old World during the time of the hunter-gatherers (the cavemen) just before the Age of Settlement. Most believe the wolf was the dog's sole ancestor, especially the Asiatic or Indian wolf (*Canis lupus pallipes*), although recent mitochondrial DNA studies suggest it was more likely the European gray wolf (*Canis l. lupus*). Others believe the coyote and the jackal may also have been involved, as they all have the same number of chromosomes (78) and produce fertile hybrids when mated with dogs. Although there have been claims, based on molecular genetics, that the wolf's domestication may have occurred much earlier, the first fossil evidence dates from about 12,000 BC, and the remains of dogs have been found in North American camp sites dating from 8400 BC. Dog bones were the most plentiful animal bones at Neolithic human sites, identified by their smaller and shorter skulls due to shortening of the jaw and

Wolf *One of the Eurasian subspecies of the wolf is generally believed to be the sole ancestor of the domestic dog. The process probably began about 12,000 BC, before man began to settle down, build shelters and grow crops. How the early hunter-gatherers (cavemen) accomplished the taming and domestication of the wolf is open to speculation, but its social (pack) life certainly aided its control.*
Photo: Geoffrey Kuchera, Shutterstock.com

smaller teeth. However, the beginnings of domestication can easily be clouded by misreading the bones. For example, the bones of foxes found in Neolithic sites in Switzerland were more numerous than dog bones, giving rise to the suggestion that the Stone Age people may have domesticated foxes as a source of food. But the red fox is a solitary and territorial animal, whose domestication has only recently been possible in the cages of fox farms, and it could not have been domesticated in a similar manner to the social wolf, as a companion animal. Apart from the dog, the only other carnivores domesticated in prehistoric times were the wild cat and the polecat, ancestor of the ferret.

The establishment of human settlements about 10,000 years ago, coupled with primitive agriculture, improved man's ability to actually keep animals, which led rather quickly to the domestication of the first food animals. From bones discovered in Iran's Kermanshah Valley in the 1970s, it is apparent that this region—on the eastern edge of the "Fertile Crescent"—was one of the first to domesticate livestock (sheep and goats), about 8000 BC. The fertile crescent extended from northern Syria south to the Nile Valley and east to the valley of the Euphrates River. Originally fertile plains and woods, as well as the cradle of civilization, it is now mainly a region of desertification due to overgrazing, deforestation, climate change, and the recent draining of the great marshes, resulting in the loss of grazing land and its wildlife. These first domesticates were traded with other areas of civilization, which became known as secondary centers of domestication. However, the domestication of these agricultural animals, more suited to the nomadic behavior of humans, may also have started before settlement began, with tamed animals following the nomadic hunters. Pigs came later, being more associated with settlements and their need for penning, or at least requiring more control than the herbivores. There are also claims that reindeer may have been herded, and their semidomesticated lifestyle started, as far back as 12,000 BC.

The wild boar, ancestor of the pig, was likely domesticated separately in two regions, firstly in China about 7000 BC, and then perhaps 1,000 years later in eastern Turkey. In the last century, wild boar were again controlled by man, this time as pure animals, when farming for their meat became fashionable. European or humpless cattle were probably first domesticated about 6000 BC in Asia Minor, Greece, or Turkey, but the humped cattle or zebus were not domesticated until 4000 BC in India. The horse and donkey were both controlled about 4000 BC, the horse in eastern Europe and the donkey in Lower Egypt. In South America, several animals were also being kept by the Andean civilizations. The guinea pig, now a popular pet animal, was first domesticated for food in the Peruvian Andes about 5000 BC; and the guanaco perhaps 1,000 years later, giving rise to the llama and alpaca long before the Inca civilization. The Incas are known to have strictly controlled llama breeding and used their meat and wool, and sacrificed males, but llamas may not have been used as beasts of burden until toward the end of the Inca period.

The yak was domesticated on the Qinghai-Tibetan plateau about 5,000 years ago, and the Bactrian camel possibly soon afterward in central Asia. The ancestry of the dromedary is uncertain, as it is known only as a domesticated animal. They were mentioned in Genesis as beasts of burden, but archaeological evidence of their use before 1000 BC is lacking. These animals, and the horse, were controlled by nomads to assist them in their travels in regions where agriculture was unprofitable. Interred cat bones dating from 7500 BC have been found in a human burial site on Cyprus, although this does not necessarily imply domestication. Prior to this discovery, it was believed that the ancient Egyptians were responsible for the wild cat's domestication, about 2000 BC. The timing of the house cat's association with humans is therefore perhaps the most uncertain of all, with a possible range of several thousand years. However, for the purposes of this book, the actual timing of the first domestication is not too important. Of more importance are the changes to wild animals that resulted from their association with developing man.

Several ancient civilizations kept birds for their symbolic value or as a source of food, and some were so suited to human control they bred readily and frequently. The rock pigeon, ancestor of the domesticated pigeon, was one of the first birds kept and regularly bred, possibly in 3000 BC in Mesopotamia, Greece, and Egypt, where it was either sacred or was a favored food. Waterfowl, both easy to control and feed, were also agreeable to total human care, but the actual origins of our modern breeds of ducks and geese are lost in antiquity. The duck was certainly one of the world's first domesticated birds, kept by the ancient Egyptians several centuries before the Christian Era, but possibly long before then in China. Newly hatched goslings that imprinted on humans made them ideal subjects for controlling, and two species of wild geese were also among the first bird domesticates. The graylag goose, ancestor of the embden goose, is believed to have been domesticated about 1500 BC; and the swan goose, which gave rise to the other main breed of domestic geese, the Chinese goose, may have been domesticated at about the same time. The Egyptian goose was kept in Egypt in 500 BC and is assumed to have been domesticated then as it breeds readily. While it did not survive as a domesticated bird, it was redomesticated again in the nineteenth century, and escapees have already established themselves as feral birds in Britain and Europe. More recently, the mute swan was kept in Britain in the twelfth century AD as a source of food, at least for ceremonial occasions. The mandarin duck, the most beautiful of all waterfowl, was first imported into Britain from China in 1745 and has bred regularly since then. The muscovy duck may have been kept by the pre-Inca civilizations, but it seems unlikely that it was domesticated by the forest Amerindians, as so often claimed; for although they are still inveterate pet keepers, with animals of all kinds in their villages, they rarely breed any of them, and certainly not from generation to generation.

Some gallinaceous or game birds were also easy to care for and bred readily. The red junglefowl, ancestor of the most important of all our food birds, the chicken, is believed to have been caged first in the Indus Valley about 2000 BC, and reached Egypt in the fifteenth century BC via sea trading routes. Kept mainly for their prolific egg-laying, they were also regarded as fertility symbols, and color mutants were recorded in Rome in the fourth century BC. The ring-necked pheasant is another example of early aviculture resulting in domestication. It may have been first controlled in China, or in Persia or Media (modern Azerbaijan and Kurdistan) but was certainly involved in ancient Greek trade and reached Rome about 2,000 years ago, for it was mentioned by Pliny the Elder (AD 23–79) in his encyclopedic *Natural History*. From there it was carried to all parts of the Roman Empire, and although there are records of breeding pheasants for the table in England in the eleventh century AD, they never assumed the same economic importance as the chicken. Domestication has produced several color mutants, as well as much hybridization between the two species and the many subspecies of the ring-necked pheasant, to the extent that most captive birds are now considered hybrids.

The blue peafowl, another Indian species domesticated there several thousand years ago, is believed to have been brought back to Europe by Alexander the Great after his campaign to the Indus River in 326 BC. They became particular favorites of the Romans and are depicted on many of their artifacts; but despite so much human control, like the ring-necked pheasant peafowl have changed only in color. The most common mutants are white, black-shouldered, and pied peafowl, but none are an improvement on the lovely wild blue bird. Guineafowl were also favorites of the early Mediterranean cultures and were eaten in Greece and Rome in 500 BC, and Caligula used them for sacrifices. Knowing how readily they breed under man's control, these quite likely were domesticated birds, presumably descendents of the subspecies of the helmeted guinea fowl that formerly occurred in Morocco. From then until the Middle Ages, however, they are missing from the records, and the current domestic stocks apparently derive from the sixteenth-century imports made by the Portuguese who traded along the West African coast.

The original or early domesticates were practically all mammals and birds. The few cold-blooded species controlled continually by man back in those days were all fishes, and the domestication of amphibians and reptiles occurred more recently. The common carp (*Cyprinus carpio*) was being raised for food in China about 3,000 years ago, and was introduced into Japan as a food fish about 1000 AD. The Crucian carp (*Carassius carassius*), ancestor of the goldfish, was also domesticated in China about 2,000 years ago. Several ancient Middle Eastern civilizations, including the Sumerians, the Ancient Egyptians, and the Assyrians, kept fish—probably tilapia, which breed very readily. In medieval times, carp were raised for food in monastery ponds and in castle and manor house moats in Europe. Keeping fish for purely ornamental purposes in glass containers began in China during the Ming dynasty (1368–1644), and in mid-eighteenth-century England the more typical aquarium containing plants and fish was developed.

Acceptance of man and his environment, ease of breeding and long-term reproduction were the classical features of the domestication process. Other desirable characteristics were the ability of animals to be tamed to reduce their flight reflexes, and an easily replaced diet. Social, group, and flock species were better adapted for such human control, as this arrangement reduced conflict at breeding time. The most useful early domesticates were those that could be controlled by tethering if necessary, such as the herbivores—the goats, cattle and horses. But these early domesticates formed just a very small segment of the animals with potential for control by man, and their numbers were limited by certain insurmountable factors of the times. Unfavorable characteristics for domestication were solitary habits and strict territorial natures. Species that met just for mating, animals that were by nature wary and nervous and had long flight distances, and those with specialized diets that were difficult to replace, were all unlikely subjects. Ancient man obviously could not cope with animals that required secure caging, such as the carnivores, crocodiles, or giant snakes.

Between the early domesticates and historic times, very few additional animals were domesticated. The Age of Discovery, beginning in the fifteenth century, saw the arrival of other species in Europe, particularly from the New World. The wild turkey was one of these, but the manner of its domestication is unclear. It arrived in Europe in the sixteenth century, and is definitely descended from the North American wild turkey (*Meleagris gallopavo*), not the Central American ocellated turkey (*M. ocellata*), which may have been kept by the Aztecs. Another historic domesticate of unclear origins is the Bengalese finch (*Lonchura domestica*). This totally controlled species, which probably derived from the white-rumped munia (*Lonchura striata*) and possibly originated in China, reached Europe in the mid-nineteenth century.

Man's close association with our two most notorious pest animals are also unclear. The Norway rat did not arrive in Europe until the sixteenth century, probably from eastern Asia, and reached England early in the eighteenth century, but the date of its domestication is unknown. It has been suggested that naturally occurring albinos may have sparked interest in caging wild rats, which were then selectively bred for their docility and fertility, eventually becoming relatively popular as pets and major research laboratory animals. Surprisingly, the rat was probably not a difficult animal to domesticate. As the results of recent experiments indicate, wild-caught rats virtually lost their territoriality and natural aggressiveness after just six generations in the laboratory. The date of the house mouse's domestication is also unknown. Color mutants can be seen in eighteenth-century paintings from China and Japan, but it is generally believed that their domestication began very much earlier.

■ THE CHANGES

Domestication is evolution and is therefore synonymous with change, and man's continual control changed animals in three major ways—in their behavior, their physiology, and their morphology. These changes resulted in most cases from the selection of existing characters, not the production of new ones. Every natural population of animals has a pool of genes hidden as recessive mutations; this is exploited by the artificial selection of animals for breeding, which is the basis for domestication. Artificial selection breaks down the established systems, and changes occur when the recessive genes are drawn out from the cover of the wild animal's genotype. Human reliance on a few genetically uniform individuals can result in genetic erosion or loss of diversity, and increases susceptibility to disease. In contrast, natural selection produces stabilized systems that ensure continuation of the wild phenotype and survival of the species, and naturally occurring genetic mutations are an integral aspect of evolution. The larger the population, the less chance of erosion.

Initially the animals controlled by man continued to resemble their ancestors in size and shape (morphologically). But they were loosely controlled and probably bred freely with their wild relatives, which happens even now for several mammals. Some domesticated gaur, banteng (only in Indonesia), yak, and water buffalo, still have opportunities to mate with wild or feral conspecifics, which is encouraged by their owners as it produces hybrid vigor—the introduction of "new blood" to counter the loss of vigor resulting from inbreeding and selective breeding for a specific trait. However, for most animals as domestication developed and they were more closely controlled and probably penned, opportunities to mate with their wild counterparts generally ceased, and mutants for color and form began to appear and to be perpetuated. This produced the distinctive initial characteristics of domestication, such as smaller body size and changes in color and pattern. This was followed eventually by selection for a specific purpose, such as greater milk or egg production, or longer wool. Economic considerations then resulted in the development of breed standards, and interbreeding with the ancestral species was then really undesirable because it could undo all the planned and selective breeding

Domesticated Turkeys *The turkey is descended from the North American turkey and not the Central American ocellated turkey. Wild turkeys were taken to Europe soon after the discovery of the New World, and it seems more likely their regular breeding and eventual domestication occurred there rather than in North America. There are records of turkeys appearing on Christmas dinner tables in Britain in 1585.*
Photo: Courtesy USDA, Agricultural Research Service

efforts. For some it was impossible anyway, as their ancestors were extinct. The outcome of producing such useful domestic animals—such as the cow and horse, for example, which served man's purpose and at the time did not appear to be improvable—meant that their wild ancestors had outlived their usefulness and some were persecuted. So the horse's ancestor, the tarpan, and the aurochs from which the European cattle are derived, were both exterminated, the tarpan mainly because wild stallions attempted to entice domestic mares away from the farm.

Although the original domestication of wild animals was therefore initially accidental, the benefits of improving animals for a specific purpose soon became quite obvious. The major interests were in the increased production of food, resulting in the control and improvement of cattle, sheep, goats, pigs, reindeer, chickens, turkeys, ducks, geese, rabbits, and guinea pigs. The value of the domesticated wolf lay in providing hunting assistance, both for game and eventually men, acting as watchdogs and companions. The polecat is believed to have been domesticated, and thus called the ferret, as a control of vermin in Rome and then eventually in British manor houses, and has only recently become a popular house pet. The role of the early domesticated cats was also as a rat and mouse catcher. The value of the horse, ass, oxen, yak, and camel as beasts of burden, and in some cases for warfare, was soon apparent; but one of the first uses of tame animals is believed to have been their employment as decoys to attract wild animals within the hunter's range, especially herd species such as reindeer and the wild cattle. The purpose of their development is shown to its extreme in the long-established domesticates—the huge size of the Ankole cow's horns, the hair length of the llama and alpaca, and the long bodies and large hams of the domestic pig. But there are also many that have no commercial applications, such as the "fancy" breeds of chickens and pigeons, with feathered legs and pompoms on their heads.

Domestication simplifies animal behavior, because foraging and hunting have been replaced with a hay rack or food dish. For some breeds, domestication results in a combination of natural behavior and learned artificial behavior—for example, the cow that grazes in its field in the manner of its wild ancestor, but then heeds the farmer's call to the milking shed where it knows there will be food in the trough. Selective breeding has reduced male aggressiveness to conspecifics, necessary in the wild but unnecessary for domesticated animals, for which dominance and hierarchy do not have important roles. There is a lack of perception of their environment in domesticated animals that is well developed in their wild ancestors.

Domestication usually restricts movement. Birds cannot migrate, and large body size and inactivity results in deterioration of the flight muscles and loss of flight. Many animals are confined in small areas, especially in intensive-breeding situations; horses and cows are tethered in stalls, dogs and elephants are chained, and camels are hobbled at night to prevent their straying. Their behavior is altered by the castration of mammals or the caponization of chickens, which reduces or eliminates aggression and has physical effects, such as increased body size, antler growth, and fat deposition. Selective breeding has also enhanced the wild canid's instinctive behavior in guarding its pack, in dogs undertaking specific roles for humans such as guarding sheep from wolves. Their natural hunting instincts have resulted in both acceptable and unacceptable domestic-influenced behavior. Hunting is acceptable (to some people) for foxhounds and otterhounds, but dogs that band together to hunt sheep are labeled killers and are usually destroyed. Following their natural instincts is less dangerous for cats, which can hunt at night and leave their catch on the doorstep the next morning. Ferrets—supposedly domesticated originally, like the cat, as vermin controls—hunt for man but with limitations. To prevent them from killing and eating the rabbits they were supposed to chase out of their burrows into purse nets, a diet of bread and milk was recommended to curb their carnivorous tendencies. Losing condition rapidly, they were then more inclined to kill rabbits just to get a decent meal, so muzzling was the answer.

An animal's physiology changes as a result of domestication. It has affected the development of the endocrine glands that are associated with reproduction, and consequently changes the estrus patterns of the wild animal. Male wolves produce sperm seasonally and breed once annually, whereas male dogs produce sperm continuously and females have two and occasionally three periods of heat each year. Alpacas and llamas breed year-round in northern zoos, while their ancestor the guanaco breeds seasonally. Domesticated animals are still reproductively compatible with their wild forebears, however, except where this is physically impossible due to size differences—for example, the chihuahua and the wolf—or because the ancestor no longer exists, such as the wild horse or tarpan, and the wild cattle or aurochs. Earlier sexual maturity in domesticated animals has also resulted from selection and improved nutrition, and the wolf breeds when almost two years old, dogs sometimes before they are nine months old. The wild animal's natural physiology is improved with selective breeding and good nutrition, resulting in increased milk production and better food conversion. Fat storage in wild animals usually occurs around the organs, especially the kidneys, whereas domesticates receiving a good diet and lacking exercise put on subcutaneous fat, plus marbling in the muscle meat.

There is greater variation of body size in domesticated animals than in their wild ancestors, with the production of both dwarf and giant breeds. Several of the first controlled animals were smaller than their ancestors, a fact so well documented that it is used to determine domesticated animals in prehistoric sites where their bones lay alongside those of wild animals. With the advent of planned breeding, any variation in body size of the offspring was exploited by man to produce new breeds, and the more recent developments have resulted in both much larger and much smaller ones in dogs, horses, cattle, chickens, and ducks. Some of the size variations affect body proportions, such as the long-bodied, short-legged bassett hound, the very long-bodied and small-headed Charolais cattle, and the short-legged Vietnamese potbellied pig, in which selective breeding has resulted in a disproportionate variation due to modification of the skeleton. However a number of animals, including the llama, reindeer, Bactrian camel, peafowl and guineafowl, have remained similar in size to their ancestors.

Inbreeding, artificial selection, and genetic drift result in changes to the wild animal's normal uniform body coloring. The loss of natural color of the wild ancestors is most noticeable between the gingery-brown wild guanaco and the black-and-white llama, in the wolf and the Dalmatian dog, and in the wild boar and the white hog. Changes in the shape of the skull, and smaller brains and smaller teeth, are among the most obvious consequences of domestication. The shortening of the facial aspect of the skull in relation to the cranial has occurred in many domesticated animals, especially dogs such as the boxer, bulldog, and Pekingese, affecting the nasal passages and making them permanently dependent on man. The Pekingese dog could not possibly survive as a feral animal, unlike many other breeds. In the domestic pig, the extreme shortening of the face has resulted in the palate bending upwards, with associated changes occurring in the cranium, and therefore in the size of the brain, and affecting the ear bones. Although the length of the bones may vary considerably between ancestor and domestic descendant, only in the tail has domestication produced any changes in the number of bones. The tail vertebrae are reduced in the Manx cat and some dog breeds, and in sheep have increased from 10 to 12 in the short-tailed breeds, and to as many as 30 in long-tailed ones. Fat storage in the tails of certain sheep breeds is a product of domestication, as this does not occur in any of the wild sheep. The digestive tract is longer in the long-domesticated carnivores, the cat, dog, and ferret, and undoubtedly reflects their increased carbohydrate consumption, as they are no longer strict carnivores. Domestication initially resulted in a reduction in horn size in sheep, cattle, and goats, but selective breeding has since more than made up for this with horns of enormous size in the Ankole cattle and Texas longhorns; four or more horns in some sheep and tightly spiraled ones in others. In contrast, several polled (hornless) domestic breeds of cattle, sheep, and goats have been produced.

■ TAMING

Wild animals have been confined by many cultures throughout the history of mankind. The ancient Egyptians were major proponents of keeping mammals and birds, and their efforts are depicted in temple and tomb art. Illustrations at Saqqara, dating from about 2500 BC, show gazelle, oryx, and ibex with collars around their necks standing in front of feed troughs. Also, hyenas are shown with pups, running with dogs and supposedly hunting with them, and lying on their backs with their legs tied together being force-fed with roasted chickens. A painting in a tomb at Thebes dating from about 1400 BC, depicting pelicans in the company of an Egyptian with two large bowls full of eggs, is said to indicate the practice of keeping pelicans in enclosures to collect their eggs. Cranes, ibis, and other birds are also depicted, the cranes being stuffed with grain to fatten them. Eastwards, in the Indus civilization of Mohenjo-Daro, the Moghul Akbar kept many cheetahs for hunting and blackbuck antelope for milking.

These examples, and others, have given rise to the assumption that some ancient civilizations, especially in Egypt, practiced experimental domestication, which they probably did, but also in claims that the depicted animals must have been domesticated. The evidence has been clearly misinterpreted, however, for knowledge of their husbandry contradicts this. The art proves only that the Egyptians kept many animals and were apparently clever at taming them. Social, herding animals, such as the gazelle and oryx, could indeed be bred in enclosures, but never in the manner they are depicted, tied in stalls like cattle. Even in recent years, hyenas have rarely bred in the best zoological gardens; and the reproductive behavior of the cheetah was only understood, and their breeding achieved, three decades ago. As pelicans lay only two eggs once yearly, a rare occurrence even in modern zoos and bird gardens, collecting eggs by the basketful could happen only at a wild colony. Domesticating a bird that needs several hundred pounds of fish annually, for the sake of a couple of eggs, seems rather futile anyway. Similarly, cranes lay only two eggs, but rarely raise more than one chick. They could not have been domesticated without a support system that included many enclosures, a better knowledge of their care, experience in sexing and artificial insemination, and certainly a higher protein diet than stuffing them with grain—none of which seems likely. However, although we will never know their intentions, we can be sure of the results. No domesticated hyenas, gazelles, oryx, ibis, cranes, or any other animals survive from those days. If they perished with the civilization that produced them, trading to secondary centers would surely have left some living evidence. It is therefore most likely that these animals were simply tamed and trained.

Elsewhere, wild animals of many species have been tamed and trained. Tame cormorants were used in China and Japan beginning in the fifth century AD for fishing, but the eggs were collected from wild nests and hatched by domestic chickens, and the chicks were hand-raised. The tethered birds, with a ring around their necks to stop them from swallowing their catch, obviously needed little training to dive for fish. Dorcas gazelle (*Gazella dorcas*) were tamed and trained by the Romans, even to the extent of pulling small chariots for children. Cheetahs were tamed by many Middle Eastern potentates for hunting gazelle. In the mid-seventeenth century, the king of Sweden's couriers rode tamed elk (the moose of Eurasia) on their duties through the forest, through which they could travel faster than horses, especially through deep snow. Moose were also trained to pull carts and were prohibited entry into the city of Tartu as they frightened horses. They were used for riding in Russia up to the end of the eighteenth century, and it has been assumed therefore that they must have been domesticated. However, although wild-born calves have occasionally been raised in northern zoos, moose have never been easy animals to keep in zoos, mainly because of their dietary requirements, and births have always been rare. It would have been impossible centuries ago to farm them to the extent of having captive-born calves for riding, let alone the continued breeding needed to qualify for domestication. The elephant is the classic example of animal taming and training. Despite being used by man for several thousand years, until recently they rarely reproduced and their captive populations consisted of tamed and trained wild-caught calves.

Some deer really have been domesticated, however. In fact, reindeer may have been the very first herbivores to be controlled by man, having been herded, and at least their semidomesticated lifestyle started, as long ago as 12,000 BC. Reindeer have been an important part of northern Eurasian culture for many years, as beasts of burden and providers of meat, milk, and hides. They are the mainstay of the Sami of Lapland and people in Russia's northlands, although many are barely controlled and still follow their migratory routes across northern Scandinavia, followed in turn by the herders who restrain them at various times for marking, gelding bulls, and culling mature animals. Others are more closely controlled as they are milked, harnessed to pull sleds, and ridden, and these animals are therefore domesticated. It is unclear whether the reindeer's control by man preceded that of the fallow deer, the only other long-domesticated deer.

■ THE VALUE OF DOMESTICATION

The long-domesticated animals have unique and important genetic backgrounds, and it is important to preserve them. But many of the original breeds—the domestic equivalent of subspecies in wild animals—have already been lost, and others are endangered. Only animals of economic interest to man are usually conserved, but even then continued selection for a particular trait may result in the loss of other potentially valuable genes; and current breeding practices favor a few breeds and their improvement. Although seemingly inferior in appearance to the large modern breeds with their thicker wool, larger udders, and rapid weight gains, the primitive breeds are a store of genetic material that has mostly been lost by modern breeding. Inbreeding, to produce the specialized desired characteristics of the farm animals of today, has reduced their genetic variability, and eventually an infusion of new genetic material must be found, not only to inject hybrid vigor, but possibly to incorporate characteristics that the primitives still possess, such as the ability to thrive on marginal lands.

The ancient breeds of animals are considered resources of unique characteristics, especially valuable for future breeding or projects to improve vigor, or to breed for a specific aspect, but they are being lost. There is therefore considerable interest in conserving ancient breeds, and the Rare Breeds Survival Trust was established in Britain in 1973 to conserve rare primitive breeds of livestock such as the Chillingham cattle, Bagot goat, and Ixworth chicken. Primitive breeds are those that have remained unchanged during the last two or three centuries of selective breeding elsewhere, and have therefore retained many of their ancestral characteristics. In 1968, at the UNESCO Biosphere Conference in Paris, member states were encouraged to preserve the genes of rare domesticated animals in view of the tendency to concentrate on selected strains. The importance of preserving genetic resources is well illustrated by the loss of size of Thailand's water buffalo. A few decades ago, bulls weighing 2,204 pounds (1,000 kg) were plentiful; but the slaughtering of such heavyweights for the meat market, instead of allowing their genes to improve the breeding stock, has reduced the national size, and even animals of 1,653 pounds (750 kg) are now uncommon

■ SOME EARLY DOMESTICATES
Wolf (*Canis lupus*): Dog (*Canis familiaris*)

The wolf originally lived throughout the northern hemisphere, from Ireland across Eurasia to Japan, south into India, and in North and Central America. Although both dog bones and those of early humans dating from several hundred thousand years ago have been discovered in the same caves, this is generally believed to be accidental due to use of the caves at different times, rather than evidence of any association between them. The wolf is usually considered to have been the first animal to be controlled by man, by cavemen about 12,000 BC; but it is regarded as accidental domestication, rather than the planned domestication of later years. Initially, it was probably a case of familiarization and accustomedness leading to taming, followed by an appreciation of their value for hunting; as cavemen had no facilities for raising or keeping wild animals, let alone the ability to maintain them long enough to breed them regularly. The wolves may have first scavenged the hunter's large mammal kills, and then perhaps became camp followers, thrown bones and offal by the caveman. Possibly they then accompanied the hunters in anticipation of sharing the kill. Their pups may well have been taken from the den and raised, which was possible if they were just weaned, but impossible for cavemen if the pups needed milk. These pups would be so accustomed to people they likely denned near their camps or caves, with their pups accepting the close proximity of man. The last stage in the process would have been active participation in the hunt, locating and tracking prey and then assisting at the kill in the manner of modern hounds. By 6000 BC dog bones were the most plentiful animal bones at Neolithic human sites; and by 4000 BC dogs were definitely assisting human hunters, driving game into nets in Egypt. The domestication of the dog preceded the cat by several millennia, the dog's social life being more suited to an association with man.

However, the ancestry of the dog is still a controversial subject. Some believe that the coyote and jackal may have also been involved, but most consider a subspecies of the Eurasian wolf the dog's sole ancestor, although which one was involved is unclear. It seems most likely that dogs reached the New World in the fashion of other mammals, across the Bering Land Bridge, but it has been suggested that the North American wolf may have been involved there separately as a dog ancestor. Breeding dogs to serve specific purposes did not begin until man progressed from an itinerant hunting and gathering lifestyle to living in more permanent communities, and years of selective breeding has produced the many modern breeds with their great range of physical characteristics and abilities. The dog shows more diversity than any other domesticated animal, and the years of control have even produced changes in voice; for barking is not well developed in wild dogs, which are generally howlers.

African Wild Cat (*Felis libyca*): House Cat (*Felis catus*)

The African or Libyan wild cat occurs as a wild animal throughout sub-Saharan Africa, excluding the rain forests. It is generally believed to be the sole ancestor of the domestic or house cat, although some biologists think that the European wild cat (*Felis sylvestris*) and possibly even the larger jungle cat (*Felis chaus*) may also have been involved. Cats are primarily solitary animals, and are less dependent upon man than the dog. In fact, they enjoy a different relationship with man than any other domesticated animal—a mutually beneficial one. The cat is provided with a home, comfort, food, companionship, and attention; while the owner has a sometimes affectionate companion whose natural hunting skills may still be practiced at night, keeping the house, barn, or granary free of rodents, or hunting small animals and thus causing considerable harm to wild populations. Very adaptable animals, cats can live a pampered life, or return to the wild and survive, either scavenging in cities or hunting in the bush.

Cats have changed little in shape and size from their ancestors despite many years of domestication. Their changes have mainly been in coat and color, and have therefore not resulted in the great variety of breeds like the dog. Their reproductive physiology has changed, however, as wild cats are monestrus—having one reproductive event annually—whereas house cats have multiple estrus periods as a result of selection and regular food availability and loss of exposure to the natural seasons. Their domestication probably began in Egypt about 4000 BC, as they depended on sedentary people—this was the time when settlement around the Nile delta began—and their first association with man may have been as mousers in the granaries. However, they may have been domesticated in Persia or Nubia centuries before, and were then brought to Eqypt. From about 1000 BC their relationship with the Egyptians changed; they were considered sacred, manifestations of the cat goddess Bastet, and were kept in temples.

Mouflon (*Ovis musimon* or *Ovis orientalis*): Sheep (*Ovis aries*)

All modern sheep are descended from the mouflon, which was probably the first food animal to be domesticated, although it may share this distinction with the wild goat (*Capra aegagrus*), ancestor of the goat. The sheep's actual ancestor may have been the European mouflon (*O. musimon*) or the Asiatic mouflon

European Mouflon   *The establishment of settlements and the beginning of primitive agriculture in the Fertile Crescent about 10,000 years ago (8000 BC) led to the domestication of the first food animals—the wild goat and the mouflon (above), ancestors of the domestic goat and sheep, respectively. Their natural herding instincts made them ideal for control by man, and sheep are believed to be the first domesticated animals selectively bred for a specific trait—the production of wool.*
Photo: Ferenc Cegledi, Shutterstock.com

(*O. orientalis*), although the former is the most likely candidate. However, there is some doubt about the European mouflon's origins, as it may itself be a feral animal derived from domesticated Asiatic mouflon brought to Europe long ago. Domesticated sheep have 54 chromosomes, the same number as the two species of mouflon. North America's dall and bighorn sheep also have 54, but this is likely just an evolutionary coincidence as they are not closely related. The first domestication of the mouflon probably occurred about 8000 BC, possibly even earlier, in the same vicinity as the goat in the Zagros Mountains of southwestern Iran bordering the Fertile Crescent, as the bones of many young sheep have been found in the Shanidar burial site there. Sometime after their domestication, they were introduced onto the Mediterranean islands of Corsica, Sardinia, Rhodes, and Cyprus, and the current, barely surviving populations there may therefore be of feral descent. Meanwhile, mouflon—perhaps pure species, intraspecific hybrids, or feral animals—still live wild throughout Europe.

Sheep are herd animals with a home range in which flocks travel in search of food, and which may vary seasonally, although they may have a definite one for the main seasons. But within their home range, they do not have territories that must be defended by the dominant male, protecting his herd. Their social structure therefore allows them to be herded into a small area, as bunching and contact is not a problem. Hardy, adaptable animals that accept human control, they have thrived in several climatic regions and, as feral animals, in many countries. The common features of today's sheep were already appearing in Mesopotamian and Babylonian art by 3000 BC, and sheep were probably the first animals to be selectively bred for a specific trait—wool production. An archeological site in Iran, dating from 4000 BC, has produced a statuette of a woolly sheep, unlike the wild mouflon that has a woolly underfur covered with heavy guard hairs. None of the wild sheep have a woolly coat comparable to that of domesticated sheep. However, hair sheep have also been developed, bred mainly for their skins, their meat, and the fact that they shed their coats, which are a mixture of hair and wool, naturally in spring and never need shearing. There are about 800 recognized breeds of sheep, many of them considered primitive, unchanged for centuries.

Tarpan (*Equus ferus*): Horse (*Equus caballus*)

The ancestry of the domestic horse is still in question. Its forbears definitely lived on the steppes of southern Russia, and they may have been tarpans. But some authorities believe the tarpan itself may not have been a wild animal, but a descendant of feral horses. It lived on the grasslands of Europe and eastern Asia, and was first described scientifically in 1774, from animals living in the Ukraine; but by then the remaining individuals were believed to be hybridized with domestic horses. In fact, Polish farmers allowed wild stallions to mate their mares, and domesticated stallions mated tarpan mares. The tarpan could not withstand this hybridization, coupled with land development, conflict with farmers, and hunting; the last wild steppe tarpan of southeastern Europe survived until the middle of the nineteenth century in the Ukraine, while the last captive horse died in the Moscow Zoo in 1875. Tarpans were small horses, about 14 hands (56 inches) high, generally dun in color with a dark dorsal stripe, mane, and tail, and a broad head with a slightly convex forehead. Domesticated horses have a long falling mane, but wild horses have a short upright one.

The domestication of the tarpan is believed to have commenced about 4000 BC by nomadic tribes on the steppes of eastern Asia. Horse molars showing bit wear, from Dereivka in the Ukraine, date from then. Horses were suited to the nomadic lifestyle of the humans of the area, who by that time already had cattle, sheep, and goats. Evidence of the horse's first use as a draft animal comes from the Sintashta fortified village site in the southern Urals, dating from 2000 BC, where horses were buried with chariots. By 1000 BC domesticated horses had spread across the grasslands of Asia and into Europe and North Africa. Heavy breeds were developed on the steppes of Asia and in Persia to carry heavily armored men. They were adopted by the Romans, and then used later to carry medieval knights into battle and tournaments, then for farmwork and pulling beer drays. Many other breeds were developed for specific purposes, including thoroughbreds for the racecourse. A selective breeding program by the brothers Heinz and Lutz Heck at the Munich Zoo, begun in the early 1930s, "recreated" tarpan using ancient horse breeds that still retained some of the tarpan genes, plus the Przewalski horse, the only remaining wild horse. Further breeding-back to produce tarpans was also conducted in Poland, and a herd of semiwild animals has been released in the Bialowieza Forest. There they joined the rare wisent or European bison, saved by captive breeding from the same fate as the tarpan.

Aurochs (*Bos primigenius*): European Cow (*Bos taurus*)

The aurochs is the ancestor of all humpless cattle, its domestication probably beginning in the Fertile Crescent of Asia Minor about 6000 BC. The European breeds of cattle are descendants of these animals. The domesticated humped cattle (*Bos indicus*), with their long dewlap and muscular hump on the withers, are believed to stem from animals that diverged from the aurochs about 200,000 years ago, and their domestication began first in India about 4000 BC.

Aurochs, as depicted on cave walls, were the wild cattle of Britain, Europe and eastern Asia. They were large black animals, the bulls standing 6 feet, 6 inches (1 m) at the shoulder, with long, lyre-shaped, forward-pointing horns that resembled the modern Spanish fighting bull. They could not compete with the growth of the human population and hunting, and became extinct in Britain in the Bronze Age, which began in 2000 BC. By the thirteenth century, they were restricted to eastern Europe, and the last animal died in 1627. As they have done with the tarpan, the Heck brothers, Heinz at Munich Zoo and Lutz at Berlin Zoo, recreated aurochs early in the last century, through back-breeding domesticated cattle breeds that had aurochs-like qualities.

Recognizable breeds of European cattle have been developed since at least 2500 BC, as man's major land suppliers of milk and meat. Their milk is the basis for many other dairy products, and they also supply fat, hides, horn, hooves, blood, bones, and dung for fuel. Cows have been used as beasts of burden in many countries since 4000 BC, and were the first pullers of wheeled transport carts. Their domestication coincided with the development of the fledgling cultivation of crops, which in northern climes provided fodder for maintaining at least the breeding stock over winter, but they were certainly more difficult to care for in such times than sheep or goats. Initially, their poor care resulted in small size compared to their wild ancestors, and the larger beef breeds were selectively bred at a later date. Changes to their physiology have resulted in the domesticated animals now being considered a distinct species; but as with all long-term domesticates, they are then subdivided into breeds, not subspecies.

Fallow Deer (*Dama dama*): Fallow Deer (*Dama dama*)

The timing and origin of the domestication of the fallow deer is lost in antiquity. But its spread across eastern Europe and Asia Minor as a controlled animal is believed to have commenced in the time of the Phoenicians, and was then spread further west by the Moors. The Normans were then apparently involved in its colonization of northern Europe, and the Romans transported deer to the British Isles. The fallow deer seen in many zoos, petting zoos, deer parks, and deer farms is therefore a domesticated animal. The natural range of the typical or European fallow deer (*Dama d. dama*) is believed to have been southern Europe and Asia Minor; but excluding the other subspecies that still barely survives in Iran, all the fallow deer now living wild and free are considered a mix of wild individuals and the descendants of feral domesticated animals. The introduced fallow deer adapted to the cooler temperate climates of western Europe and eventually became the favorite animal in British deer parks; there it is still the most common deer, existing in fenced or walled areas on hundreds of estates. It also roams freely in England's remaining forests such as Epping Forest and the New Forest, on the slopes of Snowdon, in Scotland, in many European countries, and on the island of Rhodes, where it originally occurred. On the Swedish island of Oland, a herd of fallow deer has lived in a walled hunting preserve since the sixteenth century. Fallow deer have also been successfully introduced into New Zealand, Australia, South Africa, Madagascar, Argentina, and the United States. The only surviving naturally wild population, unaffected by feral animals, is believed to be that of the Persian or Mesopotamian fallow deer (*Dama. d. mesopotamica*), which once had a wide range throughout the Middle East but was hunted almost to extinction and is now restricted to a small area of southern Iran. Its population there, and in some European zoos, totals no more than a few hundred animals. It is slightly larger than the typical fallow deer and differs also in antler formation, with less heavy palmation.

Red Junglefowl (*Gallus gallus*): Chicken (*Gallus gallus domesticus*)

The red junglefowl is the most common gallinaceous bird (chicken-like birds in the order *Galliformes*) of southern Asia, ranging from Pakistan eastwards to southern China and then south through the Malay Peninsula to Indonesia. Several subspecies evolved across this large area, but unfortunately in most places they have been exposed to chickens and feral chickens, and it is thought that few pure specimens of any subspecies now exist due to interbreeding. The pure wild bird is chicken-sized and one of the most colorful birds in the world. Males have fleshy red facial wattles, a cape of gold and bronze feathers, and a tail of long and arching iridescent green and blue feathers. It is the sole ancestor of the domesticated chicken, as the other three species of junglefowl do not produce fertile hybrids with the red junglefowl, or therefore with the chicken, and is believed to have been controlled first by humans in the Indus Valley at least 3,000 years before the Christian Era. From there chickens (now also known as fowl or poultry) spread out rapidly by trade, soon reached Persia and then Europe, and were known to the Greeks and Romans by 500 BC. They reached Britain before the Roman conquest.

Junglefowl were initially kept for their meat and eggs, but they eventually developed religious significance as a symbol of fertility due to their high rate of egg laying. The ancient Romans also used them as oracles to foretell the outcome of impending battles, and they were considered useful as alarm clocks, the male's early morning calling awakening the villagers. The first large-scale breeding—the first poultry farms—were probably developed by the Romans, who were producing color morphs in the fourth century BC. To fatten them quickly, they also stuffed them with more food than they would normally eat. The loss of flight, due to caging and their large weight gains, then allowed the breast muscles, unused for flapping flight, to develop into the favored white meat rather than the dark meat of the flying bird. To reduce aggression, thereby further improving growth rates and meat quality, the art of caponizing (castration) was apparently first devised by the ancient Greeks. The chicken is now the most common domesticated animal.

Graylag Geese   *The graylag goose is the ancestor of the 80 or so breeds of domesticated European geese. The other domesticates, the Asiatic or oriental geese, are descended from the swan goose. The graylag goose's domestication is believed to have occurred in either Greece or Egypt, probably about 1,500 years before the Christian Era.*
Photo: Jerome Whittingham, Shutterstock.com

Graylag Goose (*Anser anser*): Domestic Goose (*Anser anser domesticus*)

The graylag goose is the ancestor of one of the two main groups of domesticated geese, often called European geese to differentiate them from the Asiatic geese that are descended from the swan goose (*Anser cynoides*). The swan goose and the 20 breeds derived from it are distinguished by the knobs on top of their bills. They include the Chinese goose, as well as the African goose, which despite its name was imported into the United States from China in the nineteenth century. The breeds of domesticated graylag geese, almost 80 of them, include the Embden, a large white, meat-producing European breed; the Landes, a smaller goose developed in France for the production of foie gras; and the Pilgrim, a medium-sized American breed in which males are white and females gray. It is uncertain who was responsible for the first domestication of the graylag goose, but it likely occurred about 1500 BC, possibly in Egypt, or possibly in Greece, as it is mentioned in Homer's Odyssey. But the Romans also kept domestic geese, which they put to very extensive use. They stuffed them with bread, milk, and honey apparently to enlarge their livers, which they favored; they ate their eggs, and they also considered them sacred. They plucked their down feathers for cushions, used their quills for writing, and considered goose fat to have medicinal properties, In 390 BC the alarm calls of geese in the Temple of Juno on the Capitoline Hill saved Rome from the attacking Gauls.

Geese have never achieved the popularity of the chicken or turkey. They are gregarious birds that can be kept in flocks only until breeding time, when they pair off and separate. They lay fewer eggs than chickens, and the goslings are slow growers. Hybrids between the breeds developed from the two wild species are fertile and have resulted in some recognized new breeds.

Wild Canary (*Serinus canaria*): Canary (*Serinus canaria domesticus*)

The wild canary is a small greenish-brown bird, with just a tinge of canary-yellow on its underparts. A native of the Canary Islands, the Azores, and Madeira, it is believed to have been caged on the Canary Islands prior to their conquest by Spain in 1402. The first record of birds in Europe stems from importations by the Spanish in 1478. From then until the sixteenth century, they maintained strict control of these delightful little birds, selling only males to other European countries. The center of canary breeding moved from Spain to Italy in the seventeenth century, mainly due to the wreck of a Spanish vessel with canaries aboard on the coast of Elba; there the escaped birds became established and provided breeding stock for the Italians. Then bird breeders in Austria acquired canaries and, during the eighteenth century, shipped birds throughout Europe and to Britain. In the nineteenth century canary breeding became centered in Germany's Hartz Mountains, from where birds were even exported back to the Canary Islands.

Although they are kept throughout the world, domesticated canary breeds have changed considerably based on the commercial popularity of some breeds over others. Some countries specialized in breeding them for their song, while others bred them for color, body type, or feathering. Thus some canaries are named and classified by their shape and the geographic areas in which they were developed, while others are grouped according to their plumage, song, or color. Different countries became known for the kind they bred—for example, the Germans bred for song, while the English and French bred for coloration. The canary is a classic example of how quickly domestication can occur, for yellow and white mutants were being recorded in the sixteenth century. They are now bred in three main groups; color canaries, type canaries, and song canaries. Color canaries include various shades of yellow, red, orange, and brown. Type canaries may have crests, feathers resembling a lizard's scales, or even a humped back. The German roller is considered the finest songster, while the waterslager mimics running water.

Crucian Carp (*Carassius carassius*): Goldfish (*Carassius auratus*)

Goldfish are the most familiar pet fish, a cold-water species able to survive in a simple goldfish bowl. They are totally domesticated, an artificially produced descendant of the Crucian carp, a greenish-golden fish with red fins and a native of China, where it is a major farmed species with almost two million tons produced in 2005. Goldfish have resulted from genetic manipulation—firstly the selective breeding of the naturally appearing gold mutant, then linebreeding to perpetuate desired characteristics. This has occurred over a period of at least 2,000 years, and probably a lot longer. Several varieties of goldfish are very hardy fish, kept outdoors all winter in northern climates, when ice covers their ponds. It was recently discovered that the Crucian carp survives in the cold, deoxygenated water of its pond in midwinter by storing glycogen in its brain. Common goldfish, shubunkins, comets, and fantails have wintered outdoors beneath the ice, but lionheads, orandas, and moors are less hardy and are normally brought in for the winter when their pond cools.

The first golden fish was recorded in China in the Chin dynasty (265–420 AD), and by the Tang dynasty (618–907) goldfish were apparently well established and raised in monasteries. They were kept in outdoor ponds originally, and then eventually indoors in bowls during the Ming dynasty (1368–1644), which allowed selection for the fancy mutants that probably would not have survived the rigors of outdoor pond life. Over the years various color and form morphs were produced, the globe-eye first appearing in 1592, the celestial in 1870, and the oranda in 1893. Goldfish were exported to Japan in 1603, to Europe (Portugal) in 1611, and to the United States in 1874.

Goldfish therefore have the longest recorded history of ornamental fish, in fact of all fish, first in outdoor ponds, and then in indoor cold-water aquaria. They have been continually selected for certain features, and several of the currently available mutants stem from the late nineteenth century and the twentieth century. The comet and veiltail were bred early in the last century. The bubble eye was produced in 1908, the pompom in 1900, and the tigerhead in 1893. All the goldfish breeds have originated in China, except for the peacock tail, or jikin, and the splayed-tail goldfish, or tosakin, which were developed in Japan; the veiltail and possibly the comet in the United States; and the Bristol shubunkin in England.

Ornamental or fancy goldfish are all breeds of the same fish (the equivalent of subspecies of a wild species), with over 300 breeds recognized in China, and each breed may have many different strains of color, fin shape, or body size. Some have features that make swimming difficult and affect their vision. All the changes from the wild type of goldfish result from mutants that have been favored and selected by linebreeding, but as most morphs are recessive compared to the wild type, there is a tendency for them to revert back to their ancestral greenish-gold wild type coloration. Common goldfish are raised in large numbers as "feeder fish" for consumption by other pets, including other fish, snakes, and even large frogs. This practice is now illegal in the United Kingdom. Goldfish are also given as carnival prizes, but this practice has also been banned in some countries. Like most domesticated animals that become established in the wild again, even in countries far removed from their original habitat, goldfish have eventually reverted to their natural greenish-golden color, which has greater survival value than bright golden coloration. Goldfish now thrive in many ponds and lakes in the United States, England, New Zealand, Australia, and throughout Europe.

2   Fish

Farming fish for food has a very long history. The Chinese were raising common carp (*Cyprinus carpio*) perhaps 3,000 years ago, long before they developed goldfish from the Crucian carp (*Carassius carassius*). Even before then, in 2500 BC, the Sumerians, the Ancient Egyptians, and the Assyrians kept fish—probably the native tilapia, which breed very readily. Unlike the results of the Chinese fish-breeding, however, which can be seen today in the form of their domesticated descendents, those Middle Eastern stocks did not survive their civilization's downfall. The Ancient Romans also kept fish and even built ponds into which they channeled sea water—perhaps the first attempt to farm marine fish. Carp were also raised for food in monastery ponds, and in castle and manor house moats in medieval Europe. The carp is now a major aquaculture species farmed in many countries, with at least 30 domesticated strains produced on European fish farms, and many strains in China and Indonesia.

Common carp were introduced into Japan as food at least 1,000 years ago, and were kept mostly in the rice paddies. Although they obviously bred to maintain the stocks, the first recorded color mutants that gave rise to the koi are not mentioned until the 1820s. At the beginning of the twentieth century, therefore, there were only two known domesticated fish, the Crucian carp and the common carp, and their ornamental descendants. The goldfish and the koi are the only fish whose origins are recorded, although as with the domestication of so many other animals, the early dates may be inaccurate. Farming therefore produced not only the first examples of domestication in the fish kingdom, but also led to the development of the first ornamental fish.

Keeping fish indoors is a more recent development. During the Ming dynasty (1368–1644), the Chinese kept goldfish in bowls, and in mid-eighteenth-century England cold-water fish were kept in glass containers, but the typical aquarium was not developed until the middle of the nineteenth century. This, coincided with an understanding of the relationship between the elements, led to the "balanced aquarium," in which there was harmony between the water, plants, and animals, and the availability of gas heat allowed tropical fish to be kept. The London Zoo opened the first public aquarium in 1853.

Like all other animals, however, this situation changed dramatically in the twentieth century. The developments made then in fish breeding, especially in its latter half, produced results in a few years that rivaled the many centuries of goldfish production. Like the breeding of carp in China and Japan long ago for both food and ornamental purposes, the recent expansion of artificial fish production serves two major markets, food production and the fish hobby. The modern term for the industry is aquaculture—the controlled cultivation of aquatic animals and plants. The animals are fish and shellfish (crustaceans and mollusks), and where the fish are concerned the practice is commonly called fish farming. It includes both freshwater and marine fish, and fish of tropical or cold-water origin. Although the name is usually associated with the farming of food fish—such as salmon in seawater and catfish in freshwater—the farming of fish for the aquarium trade is also a major international industry. Facilities that raise fish for eventual release, for restocking lakes and rivers with "recreational" fish, are more properly called fish hatcheries.

Aquaculture is one of the fastest-growing food industries, with thousands of fish farms around the world, on land, and in the sea. From the food production point of view, it is promoted as the answer to providing massive quantities of animal protein at a time when the natural populations of the world's waters have been decimated by overfishing, pollution, and climate change, and many commercial fisheries have collapsed. Two major recent disasters were the collapse of the North Atlantic cod fishery and the anchovy harvest of Peruvian coastal waters. The most familiar and traditional farmed marine fish, the Atlantic salmon (*Salmo salar*), even though itself only farmed for a few decades, has now been joined by several other species. Sea bass (*Dicentrarchus labrox*), sea bream (*Saprus aurata*), and cod (*Gadus morhua*) are raised in sea cages. Two flatfish, the turbot (*Scopthalmus maximus*) and the huge Atlantic or white halibut (*Hippoglossus hippoglossus*) are also farmed, in Nova Scotia and Ireland, in land-based tanks into which seawater is pumped.

The benefits of farming fish include increased food supplies, production stability, reliability, and often reduced costs. The artificial production of fish has also reduced the demands on wild populations of both food and hobby fish. Despite these benefits, however, certain aspects are highly criticized. Although there may be some pollution of the local water table from land-based fish-farming, marine aquaculture is most often criticized for its environmental impact. Pollution from fish wastes, uneaten food, and the chemicals used to control disease and parasites, are the major concern. Escapees that hybridize with local related fish is also a problem of major proportions, with far-reaching effects. Following their hybridization with escaped farm salmon, some free-living Atlantic salmon apparently cannot navigate across the Atlantic to their traditional spawning rivers. Thousands of salmon escape annually from their cages. In May 2004, 33,000 escaped from a farm in Nootka Sound on the west coast of Vancouver Island. Preventing escapes is impossible, but there is currently interest in organic marine aquaculture in answer to the pollution concerns. Also, moving the cages several miles offshore, where they are suspended in water almost 200 feet (61 m) deep, is expected to reduce the environmental impact caused by farming in shallow tidal water. Feeding buoys, floating above the cages, are timed to release food.

Farmed freshwater fish include sturgeon, several species of catfish, carp, tilapia, trout, and bass. The amounts produced are staggering. According to the Food and Agricultural Organization (FAO) of the United Nations, aquaculture is the fastest growing food production industry with a global output of about 30 million tons. Asia dominates in the production of farmed fish, producing two-thirds of the world production. Carp are the most popular species for farming and account for about 10 million tons of the world total, most being produced in China. Tilapia is the next most popular freshwater species for farming, with 1.5 million tons produced globally—China again leading the annual production with almost half of this crop. Catfish farming in the United States is a multibillion dollar industry, with Mississippi alone producing 350 million pounds of catfish in 2005, 55 percent of the total national production. Land-based fish farms in Bangladesh produce over 1½ million tons of fish annually, mostly of the Indian carp (*Cirrhinus mrigala*), common carp, and tilapia.

The production of freshwater fish is obviously land-based. It ranges from muddy-bottomed ponds with quite murky water, to indoor commercial operations with enormous tanks for the food fish or rows of aquaria for the ornamental fish, but always with sophisticated lighting and water-quality control mechanisms for temperature, aeration, and filtration. Marine aquaculture is a fairly recent innovation, and the culture of food fish is conducted at sea in large mesh cages, as the cost of reconstituting salt water inland would be prohibitive. However, land-based farms adjacent to the sea, and a supply of clean salt water, are increasing. With few exceptions, the production of freshwater tropical fish takes place indoors in tanks in cold temperate regions, and outdoors in natural ponds in warmer climes. One exception is the outdoor fish farm in the temperate climate of southern Oregon, where cichlids are bred year-round in the geothermally heated alkaline water.

Two major production systems are involved in fish farming. When the focus is on a single species, generally of aggressive fish with carnivorous diets such as salmon and catfish, it is called monoculture. The other method, known as polyculture, involves two or more species of compatible fish. Herbivorous fish such as tilapia and carp are generally more conducive to polyculture, also termed sustainable aquaculture, in which the fish wastes act as fertilizers (rather than pollutants) for crop production. In China four species of carp may be kept in the same pond. Silver carp (*Hypophthalmichthys molitrix*) and bighead carp (*Aristicthys nobilis*) are filter feeders, eating the phytoplankton and zooplankton. The grass carp (*Ctenopharyngodon idellus*) is a vegetarian that favors pond weed, and the common carp eats detritus on the pond bottom. Each species therefore fills a particular niche, and they do not conflict. The giant gourami (*Osphronemus gouramy*), a large relative of the familiar aquarium fish the dwarf gourami as well as the Siamese fighting fish, is an important food fish in Southeast Asia, especially Indonesia. A labyrinthine fish with a specialized respiratory organ, it can survive out of water for several hours. It reaches a length of 28 inches (70 cm) and may weigh 20 pounds (9 kg). One of the most unusual fishes to be farmed (in Thailand) since the wild stocks decreased is the striped snakehead (*Channa striatus*). A predatory freshwater species, it is raised in dug-outs, net cages suspended in canals, converted paddy fields, and dammed ditches, generally in a monoculture system as it is highly carnivorous. Snakeheads are fed pellets based on a fish meal derived from "trash-fish" that are unsuitable for human consumption.

In some countries, other agricultural practices are integrated with fish production, and combining it with the farming of other animals, and crop growing, has been practiced in China for centuries. Pig manure is highly favored for fish production, fortunately indirectly. In Thailand, integrated systems combine pig and duck farming with fish farming, at least of the herbivorous tilapia, carp, and the omnivorous iridescent catfish (*Pangasius hypopthalmus*), and the manure and uneaten food provides the nutrients for the phytoplankton that the fish eat. The farm systems are integrated, livestock pens being placed alongside the fish ponds so that the manure can be washed straight into the water. Ducks, and sometimes chickens, may be housed in pens above the ponds, and their wastes drop through the mesh floors. The advantages of integrated farming are the recycling of resources, a reduction of pollution, and low cost. Fish may also be integrated with silk farms, when the silk moth pupae, which are normally discarded after the silk of the cocoon has been unraveled (except when needed as breeding stock), are used as fish food. When the ponds are cleaned, the rich silt is used to fertilize crops, which in turn feed the livestock whose manure enriches the pond water. Integrated systems are now being used in Europe and North America, with the water from the fish tanks being circulated through hydroponic beds where herbs are cultivated.

Worldwide, the trade in aquarium fish has also developed into a multibillion dollar industry. Freshwater species, especially tropicals, account for the bulk of the

Giant Gourami   *A large relative of the popular freshwater aquarium fish the dwarf gourami and the Siamese fighting fish, the giant gourami is farmed in Indonesia as a source of food. It weighs 20 pounds (9 kg) when adult, and with its specialized labyrinthine respiratory organ it thrives in shallow and muddy rice paddies, and can survive out of water for several hours.*
Photo: Courtesy Arpingstone, Wikipedia.com

trade, and the FAO considers 90 percent of all aquarium species to be captive-bred. In Asia there are major ornamental fish-breeders in Japan, Singapore, Thailand, Indonesia, Hong Kong, and the Philippines. In South America there are many in Colombia, Peru, and Brazil. Individual breeders in Europe and North America specialize in a certain group of fish or even species, producing guppies, lake cichlids, discus fish, angelfish, or gouramies. Breeding also occurs in the home aquarium, although tanks often house communal collections, where fish eggs and fry are food for others. Breeding has commenced in a few species of tropical marine fish, a very worthwhile beginning as so many tropical reefs are being destroyed or denuded. Clownfish of three species (*Amphiprion ocellaris, A. percula,* and *Premnas biaculatus*), and Barbour's seahorse (*Hippocampus barbouri*) are being bred commercially in Australia, where the freshwater Australian lungfish (*Neoceratodus forsteri*) is also raised for export.

■ FISH BREEDING AND DOMESTICATION

The rapid advances made in fish farming and breeding in recent years are comparable to the improvements made in domestic farm animal production. Numerous species have been controlled for many generations, and in most cases have been inbred, linebred or hybridized, so that many no longer resemble their ancestors in color or pattern, and in some cases not even in shape or size, and they are undoubtedly domesticated. In fact, fish are the most domesticated cold-blooded animals, with many more examples of human control and selection than of amphibians and reptiles combined.

Most controlled species, at least the food fish, are unable to breed naturally, which actually aids their captive reproduction, especially their selection for a specific trait. Most fish fertilize their eggs externally, the females laying their eggs and the males also ejecting their sperm, called milt, into the water at the same time. The conditions required by some fish, such as the open-sea egg-scatterers like cod, and the salmon that swim miles upriver to spawn, obviously cannot be reproduced in the fish farm, so breeding is achieved artificially. This involves "stripping" the eggs and milt from the fish, by gently stroking their stomachs and extruding the gametes. They are then simply mixed in a pail and hatched, and the fry are raised, all under carefully controlled conditions. These fish therefore do not actually breed in the accepted manner. Tilapia, which are mouth brooders, are generally allowed to mate and spawn naturally in their ponds, when the fertilized eggs are then held in a parent's mouth until they hatch, but they may be taken to be hatched and raised artificially. Injections of gonadotropin—a hormone from the pituitary glands of other fish—are now used to stimulate reproduction in farmed fish. Ornamental tropical fish, small species such as the angel fish and cichlids in aquaria, and the large dragon fish in outdoor ponds breed naturally. However, dragon fish eggs may be taken from the parent's mouths to hatch and raise separately, as they lay few eggs and the young fish are very valuable.

Continued breeding results in domestication, and the manipulation of captive fish through purposeful selection, and the development of mutants, hybrids, and strains, undoubtedly produces a much higher degree. The long-domesticated goldfish and koi carp are no longer the only fish to have reached this stage of artificial sophistication. The signs of advanced domestication are now quite obvious in all the major groups of fish involved in food and ornamental aquaculture. There are generally valid reasons for the production of a "new" fish. For food fish, they are improved food conversion and faster growth, and a greater tolerance to disease and water pollution. For aquarium species, it may be an interest in genetics, but more often it is the potential value of new mutants.

Continued selection and crossing has resulted in numerous hybrids and strains of tilapia, and some are identified by numbers, reminiscent of the modern poultry industry, or strains of laboratory mice. Tilapia farming was introduced into Taiwan in 1946 as an easily and cheaply produced food source. The fish currently farmed there, and in South America, are hybrids between the Nile tilapia (*Oreochromis niloticus*) and a mutant female reddish-orange Mozambique tilapia (*O. mossambicus*). Called red tilapia, they are especially favored because they resemble the popular red snapper. Light silver and white tilapia have also been developed. The fish known as hybrid striped or sunshine bass are the result of fertilizing white bass (*Morone chrysops*) eggs with striped bass (*Morone saxtilis*) sperm. This practice began in the late 1980s, and the hybrids are now a major aquaculture species. Florida largemouth bass (*Micropterus salmoides floridanus*) have been crossed with Georgia largemouth bass (*M. s. salmoides*) to produce faster-growing fish. The Georgia giant bream is a hybrid that grows much faster and larger than regular bream. Channel catfish (*Ictalurus punctatus*) and blue catfish (*I. furcatus*) have been hybridized, and have proved ideal for pond culture. They have a much faster growth rate than pure channel catfish, are more disease resistant and more tolerant of low-oxygen water, and provide a greater return on investment than channel catfish. NWAC 103, a line of channel catfish developed at the USDA Catfish Genetics Research Unit at Stoneville, Mississippi, is the fastest-growing catfish, reaching sexual maturity when two years old, a year ahead of normal.

Even more morphs have been produced by the breeders of aquarium fish. So many hybrids and color morphs of the Lake Malawi peacock cichlids (*Aulonocara*) have been bred that identifying the members of the genus is now difficult. Although there are actually only three species of discus fish, there are numerous color and pattern varieties, some of which do not breed true, so that new ones are frequently being produced. The favorite South American cichlids called oscars (*Astronotus ocellatus*) have been regularly bred by aquarists, and several color morphs are available, including the popular albino. Even more popular neotropical cichlids are the angelfish, now also being bred in a bewildering range of mutants, including zebra lace, black midnight, pearlscale, and the veiltail. Guppies, the most popular and plentiful tropical freshwater fish, and one of the least expensive, are available in a fantastic range of color and pattern morphs, differing mainly in the size and color of their fins and tails. The first genetically modified animal available as a pet—the glofish—was developed for the ornamental fish trade from the zebra fish, and is now available in three colors, red, green, and orange. These very attractive fish are fluorescent and glow at night if their tank is lit with a black light. The regular zebra fish is highly regarded as a laboratory animal, used for studying genetics and vertebrate development. Its clear eggs, developing outside the mother's body, allow the growth of the embryo to be closely monitored under the microscope.

In most domesticated animals, color and pattern mutants are generally more acceptable than morphological (shape and size) ones unless they are in proportion, such as miniature ponies, rather than the misproportioned basset hound. In fish, however, modifications to their form—primarily tails and fins—have resulted in some beautiful varieties and strains of guppies, goldfish, and Siamese fighting fish. But the production of some new fish, deformed mutants like the blood parrotfish, are ugly and malformed creatures that should never have been perpetuated. This is not the parrot cichlid (*Hoplarchus psittacus*), but an artificially produced hybrid fish available only in the last decade. It is believed to be either a cross between the midas cichlid (*Cichlasoma citrinellus*) and the red-headed cichlid (*C. synspilum*) or between the gold severum (*C. severum*) and the red devil (*C. labiatus*). It has a round, balloon-shaped body, and cannot close its mouth because of its unnatural shape, which makes eating difficult. Also, its swim bladder is deformed and affects its buoyancy. Briefly popular fish recently were the flowerhorn cichlids, a new artificial hybrid between two tropical American cichlids, *Cichlasoma trimaculatum* and *C. festae,* with the possible involvement also of the red devil, and perhaps others, the actual mix known only to the breeders. The offspring were then selectively bred to produce specimens with a large nuchal hump (on the nape) and with bolder colors and highly contrasting black markings. They were produced in Malaysia just a few years ago and were very popular when they first appeared, but it was a short-lived craze which then led to the dumping of unwanted fish in Malaysian rivers.

■ SOME OF THE SPECIES
Food Fish
Tilapia (*Cichlidae*)

Tilapia are tropical freshwater fish, natives of Africa and the Middle East, but are now established in many of the world's warmer regions. They are members of the *Cichlidae* family, well known to aquarists for its many impressive species, especially those from the African Rift Valley lakes. Three species are currently farmed—the Nile tilapia (*Oreochromis nilotica*), the blue tilapia (*O. aureus*) and the Mozambique tilapia (*O. mossambicus*). Tilapia were farmed by the Ancient Egyptians; one species, *Sarotherodon galileus,* is common in the Sea of Galilee and is likely the fish that fed the multitudes. They are laterally compressed fish that resemble perch or bass, but unlike them have a single long dorsal fin instead of two. They have large spines on the dorsal fin, also some on the pelvic and anal fins. Most tilapia are nest builders, and one of the parents guards the eggs at all times. Several species are mouth brooders that carry the eggs in their mouths for their total incubation period. The female Nile tilapia and Mozambique tilapia carry their eggs, while the male black-chinned tilapia (*Sarotherodon melanotheron*) cares for the eggs in his mouth.

Tilapia have been introduced into almost 100 countries worldwide, initially for insect and weed control, but lately for farming; they are now second to carp as the most widely farmed tropical food fish. In Asia the major farmers are China, with an

Extruding Tilapia Eggs   *On a fish farm in Brazil, eggs are extruded from the mouth of a hybrid tilapia (O. niloticus crossed with O. mossambicus) for artificial incubation. This hybrid, called the red tilapia, is favored because it resembles the popular red snapper. A mouth brooding cichlid, originally from Africa and the Middle East, the tilapia is now farmed commercially in many tropical countries.*
Photo: Courtesy University of Arizona

annual production of 700,000 tons, and the Philippines, Indonesia, and Thailand, which all produce over 100,000 tons annually. They are kept in a variety of systems, including simple muddy bottomed dug-outs, dammed irrigation channels, rice paddies, and net cages supported by bamboo poles in canals. World tilapia production in 2005 was 1.5 million tons. Taiwan exports most of its production, sending frozen tilapia to the United States and fresh fish to Japan for the sashimi market. Imports into the United States totaled 70,000 tons in the first seven months of 2005, which also included fresh tilapia fillets from Ecuador, Costa Rica, and Honduras. Tilapia farming is becoming more popular in the United States, although indoor tanks are necessary in many regions, as tilapia prefer their water at about 85° F (29.5°C). Farmed tilapia mature at the age of six months on fish farms, compared to one year in the wild, and at eight months old they weigh up to 24 ounces (680 g). They "dress-out" (cleaned with head off) at 50 percent of their live weight and at 35 percent for fillets.

Tilapia are very versatile and tolerate poor-quality water. Although basically freshwater fish, they can survive in brackish water and even seawater, and cope with low oxygen, high ammonia, and high water temperatures. But they do not like cool water, and will not feed if the temperature drops below 63°F (18.3°C).

The major factor in their favor, however, is their ability to grow on a vegetable diet (mainly algae), unlike the carnivorous salmon and trout. Consequently the natural biological production of their muddy ponds—the algae and zooplankton,which feed on the fish wastes—may provide sufficient food, although for faster growth they are usually supplemented with pellets made of soy flour. Tilapia are filter feeders, but unlike true filterers such as the silver carp—which has gill rakes that prevent solids entering its gill cavities—tilapia collect their food via the secretion of a mucus to which the algae adheres and is then swallowed. They have evolved a highly acidic stomach (with a pH of less than 2), which breaks down the cell walls of the algae, allowing the digestion of its protein, and they have also evolved fine teeth for grinding plant tissues.

Sturgeon (*Acipenseridae*)

Sturgeon are very primitive toothless fish, with long shark-like bodies covered with rows of bony plates or scutes. They are all large fish and are "bottom-feeders" that stir up the river bottom with their long snouts and locate mollusks and crustaceans with their barbels. A few of the 27 species live solely in freshwater, but most are anadromous fish, like the salmon spending part of their lives at sea and in rivers. They are found only in the colder waters of the northern hemisphere, where they have been caught for centuries for their flesh and eggs, which, when processed and salted and marketed as caviar, are the world's most expensive delicacy. Caviar from Caspian Sea sturgeon is the most prized, with the finest sevruga caviar from *Acipensor stellatus* and beluga caviar from *Huso huso,* retailing for several hundred dollars an ounce. Wild sturgeon have been decimated in most rivers, and concern for their future, and the need to ensure a supply of caviar, has resulted in the fairly new industry of sturgeon farming.

The captive breeding of sturgeon was developed in Iran and Russia, around the Caspian Sea. In Russia eight sturgeon breeding facilities in the Volga delta produced 70 million fish in four years, achievable only because sturgeon lay several hundred thousand eggs, even up to one million by a very large female. However, China has had rapid growth of sturgeon farming in the past decade, mainly of the Amur River sturgeon (*Acipensor schrenki*), but also of the Siberian sturgeon (*Acipensor baeri*) and Russian sturgeon (*A.gueldenstati*), and it may already be the world's largest producer. Outside China there are currently about 40 sturgeon farms in the world, mainly in Russia, Iran, Italy, France, Spain, Poland, Germany, and the United States, and they produce about 20 tons of caviar annually. By 2010 the global farm production is expected to be 50 tons. There are four sturgeon farms in Florida, mostly raising beluga sturgeon (*Huso huso*), and there are several in California. One of these, in the San Joaquin Valley just north of Sacramento, farms white sturgeon (*Acipensor transmontanus*), a native of San Francisco Bay and rivers of the Pacific Northwest. It is the largest anadromous fish in North America, reaching a length of 18 feet (5.5 m) and weighing up to 1,500 pounds (680 kg). The largest indoor aquaculture facility in the world is being built in Germany for sturgeon, the aim being to produce 33 tons of caviar annually. It is the first of 20 facilities planned to be built around the world in the next six years. Sturgeon grow much faster in farms. Commercial-sized fish are produced in three to four years—for their flesh, not eggs—whereas similar growth takes eight to 10 years in the wild.

Carp (*Cyprinidae*)

Carp are fish of both temperate and tropical waters, members of the *Cyprinidae,* a large family with about 2,000 species, including the goldfish and minnows. Carp are the most widely farmed of all fish, and the most productive, accounting for over one-third (10 million tons) of the world's annual farmed fish production. Most are raised in China, where the 2005 crop was reported to be about eight million tons. The silver carp, grass carp, bighead carp, and common carp are the most farmed species in China, while carp of the genera *Catla, Labeo,* and *Cirrhina* are preferred in India and Bangladesh. Carp are fast-growing fish that need little care, and thrive on the natural biological production of mud-based ponds if they are not over-stocked, although they need supplemental feeding to boost their growth rate.

Carp farming began several millennia ago, when people began to settle away from the rivers and needed a permanent supply of water. The retention ponds, and irrigation systems they dug for their crops, were also perfect for culturing carp. The eventual cultivation of rice in flooded fields or paddies allowed the now-common practice of combining agriculture with fish farming in Indonesia, Malaysia, China, and Japan. Rice can be grown dry, like grain, but the flooded field method of cultivation results in faster growth and allows at least two crops to be harvested annually. The carp "fingerlings" are added to the paddies soon after the rice is planted. The decaying leaves of the plants provide a rich source of food for algae growth, which in turn feeds the young carp. The growing fish also eat other natural foods such as insects and snails, but they may also be fed to increase the yield, which can be up to 11,000 pounds (5,000 kg) of fish per 2½ acres (1 ha). The fish in turn fertilize the rice plants with their feces, and the rich organic silt feeds the next crop of rice. The fish are harvested just before the paddies are drained to gather the rice.

Until recently, some of these farmed carp were not domestically produced, as several Asiatic carp do not breed in muddy ponds or shallow rice paddies, so "farm-ing" these fish originally involved catching the wild fry and raising them, not through breeding and domestication. The common carp and Crucian carp are exceptions, as they breed readily in ponds, often in literally stagnant water. How-ever, modern science applied to fish culture has solved the breeding problems of the other carp, and many more species will likely soon be domesticated. The artifi-cial spawning of these fish has been made possible (since 1957), by injections of the pituitary glands of other fish, which induce egg development, allowing the eggs to be stripped from the fish and then fertilized when mixed with the male's milt. Consequently, interspecific hybrids have already been produced in this manner between several species of *Labeo* carp, and intergeneric hybrids have been produced between the catla carp (*Catla catla*) and the rohu (*Labeo rohita*). An interesting side effect of this practice is that some of the induced interspecific hybrids later bred normally (naturally), and did not need to be artificially stripped and fertilized. Several species of Asiatic carp (silver, grass, big-headed, and rohu) are farmed in

Laos. They are fast growing and need little supplementary feeding if their ponds are not overstocked. The brood fish are kept in separate ponds, where they are supplemented with pellets of rice bran and peanuts, or water weeds and boiled tapioca for the highly vegetarian grass carp, and then injected with pituitary hormones to induce breeding.

Catfish (*Ictaluridae*)

There are over 2,000 species of catfish, most of which live in tropical and temperate freshwaters. They are named for their barbels that resemble cat's whiskers, although not all species have them. They range in size from the tiny candiru (*Vendelia cirrhosa*), a South American parasitic species of animals and man, to the huge Mekong River giant catfish (*Pangasianodon gigas*) that may weigh 650 pounds (295 kg). Despite their generally carnivorous eating habits—they are mainly scavengers of dead fish and other creatures, although they also eat live snails and mollusks— some catfish have proven ideal species for fish farming, and they are fed a diet of scientifically formulated high protein waterproof pellets, based on fish meal. Catfish farming is a multibillion dollar industry in the United States, the largest sector of American aquaculture, with channel, blue, white, and black bullhead species all being raised commercially. Channel catfish (*Ictalurus punctatus*) are the most popular farmed species, however. In 2005, over 600 million pounds of catfish were produced, more than all other farmed fish species combined. Most farms are in Mississippi, Alabama, Louisiana, and Arkansas, where they employ over 13,000 people and contribute billions of dollars annually to the economy. They are kept in outdoor ponds that are usually made with levees or embankments, and may be several acres in extent. Channel catfish have a breeding life of four to six years, and lay up to 4,000 eggs annually per pound of body weight, which in farmed animals usually averages 3.3 pounds (1.5 kg). Much larger wild specimens have been caught, often up to 20 pounds (9 kg).

Farmed catfish are thoroughly domesticated. Although an industry in the United States only since the middle of the last century, they have been selectively bred and hybridized, and strains have been produced. They are especially suited for intensive farming, which are enclosed systems with high fish densities; but they require water purification, regular food supplies and aeration, although catfish are atmospheric breathers and can tolerate low oxygen levels. However, as they are carnivores and do not eat algae, this must be controlled with herbicides to prevent algal blooms—when the buildup of algae followed by its death and decomposition can totally deplete a pond of its oxygen and cause high fish mortality. Channel catfish spawn from April to June when the water temperature reaches 70°F (21°C), and the males guard their clumps of adhesive eggs, which hatch in four to 10 days depending on the temperature. The hatchling catfish, called fry, absorb their yolk sacs for four days before they swim.

In tropical Asia, several species of shark catfish (*Pangasius*) are farmed in muddy ponds. The fast-growing yellowtail catfish (*P. pangasius*) and the iridescent shark or sutchi catfish (*P. hypopthalmus*) are farmed in India, Myanmar, Bangladesh, and Southeast Asia, and reach weights of up to 6½ pounds (3 kg). To induce

NWAC 103 Channel Catfish   *The channel catfish is now thoroughly domesticated, and is the most popular farmed fish in the United States. Continued selection and cross-breeding has produced many strains, and since 1986 the USDA/ARS Catfish Genetic Research Unit has been developing a line of catfish, now given the experimental name of NWAC 103. These fast-growing fish mature in two-thirds the time of typical catfish.*
Photo: Courtesy USDA, Agricultural Research Service

spawning, the brood fish (both male and female) are injected with a pituitary extract from carp. An albino mutant of the shark catfish is also bred in Thailand.

Aquaculture in sub-Saharan Africa has developed over the last four decades, especially in Nigeria, although it is still mainly small scale. The African catfish (*Clarias gariepinus*) is the favored species because of its rapid growth rate and resistance to stress, but the major control on its expansion is the growing demand for water and competition from other water users.

Salmon (*Salmonidae*)

Salmon are anadromous fish, with a combined marine and freshwater lifestyle. They spend most of their lives at sea, then migrate to rivers, sometimes travelling upstream for many miles to their traditional spawning sites. A major species in the eastern Pacific is the chum salmon (*Oncorhynchus keta*), whose life cycle is typical of the salmon. It spawns in the rivers of British Columbia and Washington in the late fall, returning from the sea and swimming upriver to where it was hatched several years earlier. Each female lays up to 3,000 eggs in the gravel, which take about four months to hatch, and the hatchlings, called alevins, remain in the gravel for about five weeks while they absorb their yolk sacs. They are then free-swimming and slowly make their way back to the ocean, migrating north to the Gulf of Alaska, where they spend the next three years. They then return to their hatching river, make a hole in the gravel with their bodies and fins and lay their eggs, which are immediately fertilized by the males, and both parents die soon afterward.

Salmon farming began in the 1960s when wild Atlantic salmon became very scarce. They are now a major food resource, with 1.2 million tons produced in 2004. The major centers of salmon farming are Norway, Scotland, British Columbia, and Chile, mostly farming Atlantic salmon which grow faster and have a greater survival rate in crowded sea pens—the large wire mesh cages, suspended over the seabed in protected bays and fiords, have been called saltwater feedlots. Practically all the Atlantic salmon now consumed are farm raised. In British Columbia, the native species of salmon—coho, sockeye, chum, and chinook (king)—are farmed, but almost 70 percent of the farmed fish are Atlantic salmon. They are also good at escaping, and escapees are now living freely in British Columbia's coastal waters, causing concern about their possible hybridization with native species and its effect on fertility and migratory habits. Breeding these fish involves stripping the eggs from the females and the milt or sperm from the males and mixing them together. The salmon hatchlings are maintained at first in freshwater tanks to simulate their natural river habitat, and then transferred to their sea pens—a very traumatic experience for them that causes heavy losses. Marine aquaculture is a highly criticized industry. Pollution is a major concern, as salmon are carnivores and need a high protein diet, often based on fish meal provided by other fish, and their feces and uneaten foods despoil the sea bed.

Ornamental Fish

Dragon Fish (*Osteoglossidae*)

The family *Osteoglossidae* contains one of the largest freshwater fish, the enormous South American arapaima (*Arapaima gigas*), and several species of the popular Asian aquarium fish known as dragon fish, members of the genus *Scleropages*. They grow up to 3 feet (90 cm) long and weigh 15½ pounds (7 kg), and their dorsal and anal fins are set far back on their bodies. Their scales are large and usually metallic and brightly colored, and they have barbels near their mouths. They live in slow-moving waters in forested swamps and wetlands, and in large fast-flowing rivers. Four species of dragon fish are recognized. The green dragon fish (*S. formosus*) lives in mainland Southeast Asian waters; the silver Asian dragon fish (*S. macrocephalus*) is a native of Borneo; the red-tailed golden dragon fish (*S. aureus*) occurs in Sumatra, and the super red dragon fish (*S. legendra*) lives in Kalimantan's Kapuas River. A fifth fish, the golden-crossback dragon fish, from Peninsula Malaysia, is believed to be a subspecies of *S. formosus*.

Dragon fish are prized aquarium fish, in great demand due to the beliefs that they are reincarnations of the mythical Chinese dragon and that they bring good luck. They are also the most expensive freshwater fish, with prize specimens, especially of the red varieties, selling for several thousand dollars. Within each species there are numerous color varieties. The super red dragon fish, with four color varieties, is the most coveted—especially the chili red with its diamond-shaped tail, and the blood red which has a fan-shaped tail. The many natural color forms have been increased by the production of captive-bred mutants.

Dragon fish are very territorial and aggressive, and are usually kept in groups in their breeding ponds, rather than pairs, as their aggression is then subdued; but they must be kept singly in aquaria. They are carnivorous; the juveniles eat insects, and the adults are fish-eaters. They are mouth brooders, and after the female has laid her eggs and the male has fertilized them, he takes them into his mouth and holds them there safely for five weeks. For such a large fish, their breeding rate is therefore rather slow, as the male can hold no more than 80 eggs in his mouth, although he normally has far less. But they do not breed in aquaria, and farmers keep them in large natural earth ponds, from which the newly hatched fry are netted and raised under more controlled conditions. The eggs are sometimes removed from the male's mouth for incubation and for raising in a safer environment. Their restricted natural populations and slow breeding rate could not keep pace with the continued demand, and to protect them they are now included in Appendix I of CITES. The demand also led to the development of breeding farms, which may also be CITES-registered, thus allowing legal trade in captive-bred fish, second filial generation and beyond. To identify these fish, they are electronically tagged with microchips when they are about 6 inches (15 cm) long.

Guppies (*Poeciliidae*)

The guppy (*Poecilia reticulata*) is the most popular tropical freshwater fish, easy to keep and breed. It is a native of northern South America, Trinidad, and Barbados; but it has been introduced and is established in many waters, including in Florida, mainly to control mosquitoes. Unwanted guppies released in Russia's cold Moscow River survive and breed near warm-water discharge pipes. There is some uncertainty about its discoverer. R.L. Guppy found them in Trinidad streams in 1866, and the species was thus named for him, but the fish were actually known to German biologists some years earlier, possibly in 1853. The wild guppy is a small fish, the males just under 1 inch (2.5 cm) long, and the females just over 1 inch (2.5 cm) long. The females are drab brown or gray, whereas the males have a colorful tail that has been greatly enhanced through selective breeding.

Early in the last century the guppy was being bred in the United Kingdom and Germany, and fish were imported into the United States in 1910. By the middle of the century, societies had been formed and guppy shows were held. The guppy is undoubtedly domesticated; a tremendous variety of shapes and colors have resulted from a century of selective breeding, and standards have been developed for many of the forms. Although most fish lay eggs that are fertilized after they are laid, the guppies—and several other aquarium favorites including the mollies, platies, and swordtails—are live bearers. This involves internal fertilization, achieved by means of a modified anal fin acting as a sexual organ. The eggs develop inside the mother and hatch there, and she gives birth to up to 60 live young after a gestation period of 28 days. In communal tanks, where they are usually kept, the tiny fry must have plenty of hiding space or they become food for other fish. Even their parents may eat them, and special tanks are available in which the fry can avoid their mother. Specialist breeders have developed hundreds of strains, affecting the guppies' color and size and the shape of their tails. Their colors include blue, green, yellow, and red, with males being the most colorful. There are unicolor guppies, which are all one color including their tails, and there are tuxedo guppies, in which the rear half of the body is black and the tail another color. Snakeskin guppies have wavy lines along their bodies that extend into the tails. The tail is certainly the domesticated guppy's most glorious aspect, and many varieties have been produced with enhanced tail size, shape, and color. There are double swordtails, lyretails, delta-tails, fantails, crowntails, and many others, all brilliantly colored and usually with dark spots.

Cichlids (*Cichlidae*)

Cichlids belong to the family *Cichlidae,* a group of fish containing about 2,500 species, that includes some very important food fish such as the tilapia, and many beautiful aquarium species, including the lake cichlids, discus fish, oscars and angel fish. They are mainly tropical freshwater fish, with a few brackish species, with a range that includes the warmer parts of the New World, from the southern United States through Central and South America; plus the Middle East, Africa and India. They vary in shape from the laterally compressed discus fish of the Amazon basin, to the very elongated carnivorous pike cichlids, also from the Neotropics.

Discus fish are a favorite aquarium species, large and colorful, and available in a wide range of color and pattern morphs. There are only three wild species—the red discus (*Symphysodon discus*) the blue discus (*S. aequifasciatus*) and a recently discovered one named *Symphysoson tarzoo*—but there are many color strains in the wild, and red and blue discus have been hybridized in the aquarium to produce many more mutants. One of these, the pigeon blood discus, was developed from a single fish in the 1980s. A dominant gene affects the distribution of black pigment, so the normal vertical black bars are missing, making the fish appear much brighter and lighter. Consequently, they cannot darken at will, like normally colored discus fish when they are stressed or when spawning—when their fry are attracted to the parent's dark skin to feed.

Angel fish, also from rivers in the Amazon basin, are laterally compressed with round bodies and elongated triangular dorsal fins. The most frequently seen wild angel is *Pterophyllum scalare,* a silver fish with three dark-brown stripes; but years of selective breeding since early in the last century, when they were first available, have resulted in many color and pattern mutants, in some of which the dark stripes are absent.

There are numerous lakes in Africa's Great Rift Valley, but three—Lakes Victoria, Tangayika, and Malawi, which lie in the Western or Albertine Rift—are very important in the aquarium fish trade, as they are home to at least 1,500 species of cichlids. These fish have been exported in large numbers, and still are, but many species breed readily and have been hybridized and selectively bred. The Malawi butterfly cichlid (*Aulonocara jacobfriebergi*), a beautiful fish that grows up to 8 inches (20 cm) long, has a number of natural color forms. It was first exported from Lake Malawi about three decades ago and has bred regularly, with several captive morphs now available, including albinos. The most beautiful mutant, however, is undoubtedly the eureka red cichlid, in which linebreeding has resulted in "fixing" and improving a desired trait, such as red fins and tail. Another favorite cichlid that has been available for many years as a captive-bred fish is the kribensis or purple cichlid (*Pelvicachromis pulcher*), of which many color morphs and albinos have been produced. The orange cockatoo cichlid (*Apistogramma cacatuoides*) also has several captive-bred mutants.

Gouramies (*Belontiidae*)

The family *Belontiidae* contains about 100 species of gouramis and related labyrinth fishes, so named because they have a special respiratory organ, and can gulp air and use atmospheric oxygen. This allows them to live in shallow and warm oxygen-depleted water, such as paddy fields and muddy irrigation ditches. The family includes many well-known aquarium fish such as the dwarf gourami, pygmy gourami, three-spot gourami, the paradise fish, and the giant gourami (*Osphronemus gouramy*) that is farmed in Indonesia for food. The related kissing gourami (*Helostoma temmincki*), with its highly protrusible lips and horny teeth, is the sole member of the family *Helostomatidae*. Many species of gouramis breed regularly in the home aquarium and in the larger facilities that produce tropical fish for the wholesale market, and there is no doubt about their domestication. The three-spot gourami (*Trichogaster trichopteris*) is now available in several color morphs, including blue and gold. The kissing gourami, which has two natural color mutants—green and rosy pink—is now also available as an artificial dwarf mutant of the latter. The paradise fish (*Macropodus opercularis*), is said to have been the first tropical fish to be kept in Europe, and was named by Linnaeus in 1758. For a tropical species it is quite hardy, able to survive in water at 55°F (12.7°C), but it was still quite a feat to bring such fish back alive by sea from eastern Asia, and then maintain them. More recently it has been bred continuously, and both albino and black morphs have been perpetuated.

The 50 or so species of "fighting fish" of the genus *Betta* are also members of this family. They are separated into two groups, depending on whether they build bubble-nests for their eggs or if they are mouth brooders. The most familiar member is the Siamese fighting fish (*Betta splendens*), one of the most popular of all freshwater tropicals. Known simply as bettas, they rival the guppy in the number of color and form varieties that have been produced. The wild betta is a greenish-brown fish with relatively short fins, but selective breeding has produced a range of brilliant colors and long fins and flowing tails in the males, and colored females recently also. Bettas are available in solid colors, the most common being red and blue, also in metallic copper and gold; and in pattern morphs that include marble and butterfly. Their tails are almost as variable as the guppies, with spade tails, round tails, delta tails, and the most common form, the veil tail. In the recently bred crown tail, the edges of the tail are fringed.

3 Amphibians

The class *Amphibia* is a very large group of vertebrate animals, with over 6,000 species of frogs, toads, salamanders, and caecilians. Yet it has fewer recently domesticated species than all the other vertebrate classes. This is simply because domestication is achieved only as a result of continued control and breeding, and until quite recently most amphibians did not breed often enough. The only exceptions were the axolotl (*Ambystoma mexicanum*), the African clawed toad (*Xenopus laevis*) and a few species of newts, notably all aquatic species. Their domestication process, which commenced in the nineteenth century for the axolotl and early in the last century for the other species, resulted solely from their usefulness to the research community. More recently, other species have shown they are also capable of sustained captive reproduction, with sufficient regularity for them to be considered either in the early stages of domestication, or with the potential to eventually become as domesticated as the axolotl and clawed toad. Unlike the first controlled amphibians, whose domestication resulted from the needs of the laboratory, these new domesticates stem mainly from the involvement of private hobbyists and zoological institutions. Also, they are not restricted to aquatic species, but include terrestrial and even arboreal forms. One other element, frog farming, has also been involved in large-scale amphibian breeding in recent years, but mainly of just one species—the bullfrog.

The deteriorating state of the global environment and its effect upon wildlife has resulted in concern for amphibians and a call for increased captive-breeding efforts. Large-scale projects have been urged for many more species in view of the threats they now face. Almost one-third of the known species are listed as critically endangered, endangered, or vulnerable by IUCN, due to the loss of habitat, pollution, drought, viral disease, increased UV-B radiation, and especially the frog fungus *Batrachochytrium dendrobatidi*. One hundred and twenty-five amphibian species have been lost since 1980. But most amphibians do not breed readily in artificial conditions, and success requires careful duplication of the smallest details of their life history and environmental requirements. This is especially true of the terrestrial species, as well as of the land-dwellers that return to the water to lay their eggs, as most species do. Many require sophisticated considerations such as manipulating the temperature to simulate their natural daily and seasonal changes, including a long-term cooling period to reproduce winter conditions; and artificial rain chambers, in which a warm mist duplicates the rainy season. The complicated reproductive strategies of many species, such as carrying their eggs or even their tadpoles on their backs, or feeding their tadpoles with eggs laid just for that purpose, add to the breeding difficulties. Consequently, amphibians have never achieved the popularity of the other vertebrates, or even of the other cold-blooded creatures, the fish and reptiles. This is very noticeable in zoos, where aquaria and reptile houses (which may include a few amphibians) far outnumber the displays now dedicated to amphibians. Their reclusive and mostly nocturnal nature is usually cited as the major reason for this; yet many of the truly aquatic forms are just as visible as fish in their tanks, and many frogs are as colorful as tropical fish.

In contrast to their originally perceived poor value for display and low interest in breeding them, amphibians were highly regarded for research. Numerous species, especially frogs and salamanders, have been widely used in the classroom for teaching and dissection, and in the laboratory for more sophisticated research, including advanced embryology, human pregnancy testing, the analysis of gonadotropins, and transplant research. But this does not necessarily mean they were captive-bred, and even now most amphibians used in research are still wild-caught. The two most adaptable species for laboratory life, the axolotl and the clawed frog, have been joined more recently by the dwarf clawed frog (*Hymenochirus boettgeri*), the Spanish ribbed newt (*Pluerodeles waltli*), alpine newt (*Triturus alpestris*), smooth newt (*T. vulgaris*), palmate newt (*T. helveticus*), and the Japanese newt (*Cynops pyrrhogasater*). These have all been maintained in the laboratory for many generations. The most significant factor of their lifestyle is their aquatic nature—either fully aquatic like the axolotl and clawed frog, or the newts that can adopt a permanently aquatic way of life. Other amphibians, such as the mainly aquatic oriental fire-bellied toad (*Bombina orientalis*) and the terrestrial tiger salamander (*Ambystoma tigrinum*) and fire salamander (*Salamandra salamandra*), are also laboratory animals. However, although these species have bred, they are not yet maintained in self-sustaining colonies, and the research requirements are largely supported by animals collected in the wild.

There are several advantages in keeping aquatic amphibians. They generally lay many more eggs than the terrestrial ones. The axolotl lays several hundred eggs, and the clawed frog several thousand, in a year. The newts are less prolific, but still lay from several dozen to 100 eggs, depending on the species. For the aquatic species, the quality of their water (the pH, temperature, filtration, and aeration) is much easier to control than the combination of wet and dry environment required by the terrestrial amphibians, with their variable air temperature between day and night, winter and summer, seasonal variations in humidity, and substrate suitability. Amphibians in water are also easier to feed, as they generally accept inanimate foods, even commercial fish food; whereas most land amphibians are usually only interested in prey that moves and thus activates their feeding response. In addition to relying upon natural breeding, the reproductive process has been initiated by injecting amphibians with synthetic hormones that encourage the production of gonadotropin, which stimulates the ovaries and testes.

Farming is the second element of amphibian husbandry that has resulted in the sustained breeding of large numbers of frogs, but of one species in particular, the bullfrog (*Rana catesbiana*). Others may be involved, in Asia at least, especially the Chinese edible or Taiwan frog (*Hoplobatrachus rugulosus*), and the Indian bullfrog (*H. tigerinus*). There are commercial bullfrog farms in Southeast Asia, Indonesia, and in the New World tropics. The major market is for frog's legs, followed by live frogs for research and education. Their large back legs, which comprise up to 40 percent of a frog's weight, are marketed for human consumption, and their skins are used to makes shoes, belts, purses, and wallets.

There is an enormous international trade in frog's legs, and farming is helping to reduce the demand on wild populations. Europe imports 6,000 tons of frog's legs annually, mainly from wild-caught frogs in Indonesia, India, and Bangladesh. But the global demand is so great that all the captive-breeding efforts currently produce only 15 percent of the annual requirement. Taiwan, the world's leading bullfrog farmer, exported 1,550 tons of frog's legs in 2002. There is also substantial production in Thailand, Brazil, Mexico, Guatemala, and Ecuador, and a farm near Montevideo produces 30,000 frogs annually for the international restaurant trade. Normally considered difficult to feed, unless stimulated by the movement of their prey, the farmed frogs are so accustomed to their artificial environment they have no such aversions, and in Taiwan bullfrogs readily accept food pellets formulated especially for them. In Uruguay they are fed trout chow, as well as cultured earthworms.

The other major amphibian breeders in recent years, mainly in the last two or three decades, have been zoos and private herpetologists. They now far exceed the research element in the production of captive-bred species, and some in numbers also. They are therefore the major contributors to the domestication of amphibians. Species that have reproduced regularly are the horned frog (*Ceratophys ornata*), White's tree frog (*Litoria caerulea*), the African bullfrog (*Pyxicephalus adspersus*), and the tomato frog (*Dyscophus antongilli*). Several of the highly colorful Madagascar mantellas have been bred, including the golden mantella (*Mantella aurantiaca*), the green mantella (*M. viridis*) and the green climbing mantella (*M. laevigata*). The tiny poison-dart frogs now have a great following in herpetoculture, both as a hobby and a commercial activity supplying hobbyists. Species that are now breeding regularly include the blue poison-dart frog (*Dendrobates azureus*), the green and black poison-dart frog (*D. auratus*), the strawberry poison-dart frog (*D. pumilio*), the splash-backed poison-dart frog (*D. galactonotus*), the dyeing poison-dart frog (*D. tinctorius*), and the golden poison-dart frog (*Phyllobates terribilis*). Several species are available in different captive-raised bloodlines.

The captive breeding of many other species has been prompted by their global plight, and breeding projects have been initiated in the amphibian's countries of origin, generally in association with the government or its agencies. These projects

Blue Poison-dart Frog *This tiny frog is now being captive-bred quite regularly, despite its very complicated biology. The male guards the eggs on the forest floor and moistens them with urine. When the tadpoles hatch, he carries them to a phytotelmata (a small pool of water in a plant axil) in the forest canopy. The female then lays eggs for them to eat.*
Photo: John Arnold, Shutterstock.com

are primarily concerned with supporting wild populations and with restocking areas where the species is endangered or has already been exterminated. Although the emphasis is therefore on the production of animals for release, the self-sustaining breeding stock of several species has been maintained for many generations. A classic example of such programs is the American Zoo and Aquarium Association's (AZA) Species Survival Plan (SSP) for the endemic Puerto Rico crested toad (*Peltophryne lemur*), which was near extinction. With many member zoos cooperating in the breeding program, 90,000 tadpoles were released into their native habitat between 1987 and 2005. Another success involves the endemic Mallorcan midwife toad (*Alytes muletensis*) saved from almost certain extinction through captive breeding projects initiated by Jersey's Durrell Wildlife Conservation Trust. Many other species are now benefitting from similar programs, including the Panamanian golden frog (*Atelopus zeteki*), the green and golden bell frog (*Litoria aurea*), the emperor newt (*Tylototriton verruscosus*), and the two very rare giant salamanders. The Japanese giant salamander (*Andrias japonicus*) has been the subject of intense breeding efforts at Hiroshima's Asa Zoological Park since 1979, and thousands of larvae have been produced for release into their traditional mountain streams. In the Chinese province of Jiangxi, a facility has been established to breed the Chinese giant salamander (*Andrias davidianus*); by 2005 it was producing 10,000 young salamanders for release annually. Even with their complicated breeding biology, the captive reproduction of these species provides further proof that most animals will breed in response to serious efforts.

■ SOME OF THE SPECIES

Axolotl (*Ambystoma mexicanum*)

Axolotls are large salamanders, about 12 inches (30 cm) long, that occur naturally in shallow lakes, such as Xochimilco and Chalco, near Mexico City. They were apparently well known to the Aztecs, who named them for their god Xolotl. They are normally totally aquatic, and with their typical dark mottled black-and-brown coloration they resemble tiger salamanders with gills. Under normal circumstances, axolotls retain their tadpole features—their feathery external gills and tails—even when they are adult and breeding, a characteristic known as neoteny. If their ponds dry out, however, axolotls may metamorphose to a terrestrial way of life like some newts, losing their gills and living on land. Pet axolotls rarely have the opportunity to metamorphose and remain aquatic animals all their lives, but laboratory specimens have been induced to become terrestrial with injections of a pituitary gland extract.

The axolotl has been a laboratory animal since the late nineteenth century, and is considered a thoroughly domesticated species. Selective breeding has produced several mutants and many strains. Albinos cannot synthesize melanin and consequently lack pigment cells, and are mottled yellow and pink with red gills and eyes. Leucistic axolotls, with dark eyes and off-white coloring, are also bred, and occasionally mottled ones. Like most domesticated species there are many more under human control than ever existed in the wild. With their unique lifestyle and regenerative powers, axolotls are both ideal laboratory animals and aquarium favorites, and are bred in large numbers. They are widely used in research, as they can regenerate amputated limbs, tail, and even external gills, and then heal without scarring. They accept limb transplants from other axolotls and can restore them to full functioning; they have even repaired and restored a damaged limb while growing an additional transplanted one. The Ambystoma Genetic Stock Center at the University of Kentucky maintains a huge self-sustaining breeding colony of axoltls and supplies biology research programs and classrooms throughout the United States and globally.

The axolotl is the classic amphibian example of rapid domestication, a process aided by its aquatic lifestyle and ease of feeding and breeding—which can commence at the age of 15 months. They may lay over 1,000 eggs at a time, although the normal clutch is about 500. Like the related newts and other salamanders, the fertilization of the axolotl is internal, unlike most frogs and toads in which the spawn is fertilized by the male after it has been laid. The male axolotl deposits a spermatophore—a cone-shaped mass of jelly and spermatozoa—that sticks to a submerged rock or log. The female then lowers herself onto it and picks it up in her cloaca—the single urogenital and anal opening also common to reptiles and birds.

Clawed Frog (*Xenopus laevis*)

The clawed frog, also called the clawed toad, is a wedge-shaped animal with a small head and eyes on top of its head. It is usually brown above with a pale belly, but albino morphs are also available. It reaches a length of 4 inches (10 cm), and females are usually slightly larger than males. This frog has a habit of floating motionless below the surface for long periods—in contrast to its energetic amplexus behavior, when it swims in circles up to the surface, where the females lay their eggs, and the males fertilize them as they are laid. A carnivore and a scavenger, it digs invertebrates out of the mud with its long claws. A member of the *Pipidae,* an entirely aquatic family, the clawed frog is a native of the cooler and higher regions of sub-Saharan Africa. There it occupies natural waters and man-made reservoirs, ditches, and wells, where it avoids drought by burrowing into the mud. It does not like the humid lowlands and is absent from the entire Congo Basin.

Highly adaptable and quite salt-tolerant, the clawed frog has been widely dispersed globally, its ability to lay several thousand eggs annually allowing it to quickly become established. Released frogs and escapees from laboratories are now pests in many countries and difficult to eradicate. In the United States they occupy several states, especially California, where 400,000 were caught in a San Joaquin reservoir recently. It is also an established alien in Chile, England, Germany, and Holland. The clawed frog is the second most domesticated amphibian after the axolotl, with a long and continuing history as a laboratory animal and as a pet. It breeds readily and large numbers have been produced for research, beginning in the first half of the twentieth century, followed by a spin-off pet trade in the second half. Its original value to science was in pregnancy testing. When injected with human urine, the frogs lay eggs if the urine contains chorionic gonadotropin hormone—an indication of pregnancy. Clawed frogs are still the most widely used research animals for cell and molecular biology, and antibiotics recently discovered in their skins are now being developed. A related species, the dwarf clawed frog (*Hymenochirus boettgeri*) of West Africa, only 1½ inches (4 cm) long and fully aquatic, is also a laboratory animal and a favorite species for the home aquarium. It is bred in Germany, Poland, the Czech Republic, and Indonesia.

White's Tree Frog (*Litoria caerulea*)

One of the 700 members of the tree frog family *Hylidae,* White's tree frog grows to about 4.5 inches (11.5 cm) long, and is a very plump animal with a smooth, pale green skin. Males have gray throats and the females are whitish. They usually have a number of small white spots on their bodies and legs, and can change color to brown. There are two subspecies of this frog, one from Indonesia, the other from Australia; and a blue morph has been produced through selective breeding, a very attractive animal in which the intensity of its color deepens with age.

White's tree frogs were named for John White (1756–1832), surgeon-general to the First Fleet and then to the new settlement eventually named Sydney, who mentioned it in his *Journal of a Voyage to New South Wales* (1790), one of the earliest accounts of Australian animals. They are adapted for life in quite arid regions, where they are protected from desiccation by their waxy skin covering and their nocturnal habits. But they are quite sedentary creatures, and seldom move far at night if food is available. Despite their inactivity, they have become one of the most popular exotic pets, as they are easy to feed, become quite tame, and may live for 20 years, so they are perfect frogs for the beginner herpetologist.

Breeders usually keep a pair in a 20-gallon (75 L) glass aquarium, covered with a fine mesh screen. The tank is decorated with house plants that may be left in their pots, and dead branches for the frogs to climb, as they are arboreal and spend most of their time above ground. As they evolved in the tropics, they need a high temperature, at least 75°–85°F (23.8°–29.4°C) during the day—which can be provided by a low-wattage lamp bulb over their tank—and 70°–75°F (21°–24°C) at night. Herpetologists feed crickets, earthworms, and mealworms to their frogs but are careful not to overfeed them, or they overeat and become obese. In their natural habitat, White's tree frogs breed in water, the female laying up to 300 eggs on several occasions while in amplexus, which may last for two days. The eggs are fertilized in the water by the male's ejected sperm. Captive frogs have bred regularly, following a routine that simulates their natural environment. This begins with a period of estivation (the summer equivalent of hibernation) when their temperature is lowered to 65°F (18.3°C) for six weeks, and the humidity is reduced, when they become semidormant. They are then moved to another tank containing water at least 8 inches (20 cm) deep, and with branches on which they can climb out of the water if they wish. They are misted with warm water several times daily to simulate spring rains, and then lay their eggs, which sink and become attached to the bottom of the tank or the submerged sections of the branches. The parents are then removed from the breeding tank. The eggs hatch after 36 hours, and the hatchling tadpoles are fed fish flakes and chopped tubifex worms.

Poison-dart Frogs (*Dendrobates*)

The poison-dart frogs (or Dendrobatids) are tiny, very colorful frogs from the rain forests of Central America and northern South America. Their color serves as

White's Tree Frog *This plump and shiny Australian frog lives in quite arid regions, where its waxy skin covering prevents dehydration. It is one of the most popular pet species, as it is easy to feed and becomes quite tame, and is now bred regularly by herpetologists. It is also very long-lived, with a potential lifespan of 20 years.*
Photo: Courtesy LiquidGhoul, Wikipedia.com

a warning to potential predators that their skin glands secrete a highly toxic substance synthesized from the invertebrates that form their diet. These insects acquired the toxins from the alkaloids that many plants produce to protect their leaves from insect attack. The Choco Indians of western Colombia "grill" the frogs on skewers over their fires and then wipe the poisonous exudate onto their darts. The poison is very effective, causing convulsions, paralysis, and the eventual death of their prey, mainly monkeys and large birds like the curassow and guan. Although poison-dart frogs were kept early in the last century, serious attempts to maintain and breed them did not commence until the 1970s. The success of these efforts is now obvious in the increasing number of captive-raised specimens being offered for sale by hobbyists and commercial breeders. But this was not easily achieved, as their breeding biology is quite complicated.

Poison-dart frogs are mainly terrestrial, and unlike most amphibians are diurnal. They lay fewer eggs than most frogs due to the more intensive care they provide for them and the ensuing tadpoles. The female lays her eggs, usually no more than six, in a moist place on the forest floor; and after he has fertilized them, the male stands guard until they hatch, moistening them with urine to keep them hydrated. He then carries the tiny, slippery tadpoles on his back to the nearest water, which is usually within the leaf axil of a bromeliad high in the forest canopy, where the small amount it contains forms a unique environment called a phytotelmata.

In several species, including the strawberry poison-dart frog (*Dendrobates pumilio*), the granular poison-dart frog (*D. granuliferus*), and the Panamanian poison-dart frog (*D. speciosus*), the mother frog cares for the eggs and carries the tadpoles to water. She then visits them regularly to feed them with eggs that she lays in each tiny pond expressly as food. In some species, the tadpoles are carnivorous and must be limited to one per pond. Intensive efforts to duplicate the frog's environment have resulted in the successful breeding and raising of several species. For a breeding pair of poison-dart frogs, a 12-gallon (45L) glass aquarium is outfitted as a vivarium, in which the equivalent of a mini-tropical conservatory is then created, with soil, moss, dead branches, pieces of bark, and living house plants, sometimes even small bromeliads, thus duplicating their natural environment. The top is covered with fly screen to prevent the frogs and their food from escaping, and a humid tropical environment is created with a daytime temperature of about 80°F (26.6° C), dropping to 73°F (23°C) at night. High humidity is essential, and is normally achieved through watering the plants, keeping the substrate damp, and misting the plants with warm water daily, but it must not be "dripping wet" all the time.

Breeders provide their frogs with upturned earthenware flowerpots or half coconut shells, with an entry hole in the side, and a petri dish inside in which they lay their eggs. These are usually removed immediately and incubated in a separate container, with just ¼ inch (6 mm) depth of water added when the eggs hatch. The tiny tadpoles are given daphia (freshwater crustaceans or waterfleas), spirulina, mosquito larvae, and flaked fish food in minute amounts, taking great care not to pollute their water. These captive-bred and artificially fed frogs are no longer "poison"-dart frogs however, as their artificial foods lack the toxins that their wild ancestors synthesized from their insect diet.

Argentine Ornate Horned Frog (*Ceratophrys ornata*)

A nocturnal, grassland species from Argentina, Uruguay and southern Brazil, the ornate horned frog, also called the pac-man frog, is now a common house pet and is bred commercially for the pet trade. It is a very impressive animal, the females reaching 9 inches (23 cm) in length, while the males are slightly smaller. They are very colorful, their dark-green base color being liberally blotched with red and black. They have squat and round bodies with enormously wide mouths, and are one of the few frogs to possess teeth. Although the other horned frogs have horns—actually projections of skin over the eyes—this species is hornless.

Horned frogs are very carnivorous animals and, with their great appetites and sedentary lifestyle, high-risk cases for obesity. Large frogs eat smaller frogs, and they are such aggressive feeders they must be kept on their own except for breeding; even then cannibalism is possible if the female is considerably larger than the male. They are ambushers, usually content to sit and wait until the next meal appears within reach, so they do not need a large vivarium. A 15-gallon (56 L) glass tank is adequate for one frog, with a substrate of soil and moss and decorated with house plants, dead branches, and large pieces of bark. A large, shallow dish is provided, with just enough depth of water for the frog to sit in without being covered, as obese individuals have drowned in deep water. They do not need the high temperatures of the tropical rainforest frogs, and are usually kept at about 73°–78°F (22.7°–25.5°C) during the day, and 68°F (20°C) at night. Adult frogs are easy to feed as they attempt to eat anything that moves, even fingers, but they are generally given mealworms, large crickets, small fish, earthworms, and young mice.

Horned frogs have been bred by simulating winter. This is achieved by moving them individually into a tank containing sphagnum moss, which is allowed to dry out. They remain there in a temperature of about 70°F (21°C) for a period of six weeks, during which time they are not fed but are given a bowl of water in case they need to drink. They are then moved in pairs to another tank containing 1 inch (2.5 cm) depth of water, and are sprayed with warm water several times daily. Plastic water lily leaves, on which the eggs will be laid, are floated on the surface. The tadpoles are also very carnivorous and are usually kept singly in jars, which entails a lot of work as they may number in the hundreds. If they are kept in a communal tank, the surface is liberally covered with floating plastic plants in which they can hide. When the tadpoles begin to develop their legs at the age of one month, thin pieces of wood are floated on the surface for them to clamber onto. Horned frogs have also been induced to breed with hormone injections, but many breeders shun this method, believing it indicates poor husbandry.

Bullfrog (*Rana catesbiana*)

The largest North American frog, the bullfrog reaches a length of 8 inches (20 cm) and weighs up to 28 ounces (800 g). It is very variable in color, from brown to green with spots or blotches on the back. Sex is determined externally by the size of the frog's tympanum or external eardrum, which is much larger than the eye in males and smaller than the eye in females. It is distributed from New Brunswick in eastern Canada south through the eastern United States to Texas and Florida;

but it has been introduced into many countries, including several in Central and South America, a number of Caribbean islands, and throughout Southeast Asia.

Historically, frog's legs have been a delicacy in several countries, and until recently were all provided by wild frogs. The bullfrog was involved in that trade, and was in fact the largest frog in the global restaurant trade, (the larger "mountain chicken" frog of Dominica is used locally) although most of the frog's legs came from smaller species harvested in Europe and India. There is still a large trade in wild frog's legs, but many are now also produced by frog farms, and the bullfrog is the major species involved. Bullfrog farming began in Mexico in the 1960s and was then taken up by several Asian countries, especially Taiwan and Thailand. Commercial breeding has been attempted in the United States, but has not been successful, probably because of the short (although natural) breeding season of four months, whereas in the tropics they breed year-round.

Depending on the country's preference, bullfrog breeding facilities may be outdoors or indoors. Outdoor pens may incorporate the "wet system," in which the frogs are kept in shallow water in concrete pens with smooth walls about 39 inches (1 m) high, with fresh water continually flowing through the pens. Or they may be kept on the "dry system," when most of their pen floor area is grassed, with shallow concrete ponds providing the water in which they mate and spawn. The breeding tanks may be covered with cloth to provide shade, or mesh-covered to keep predators out. In the controlled environment of indoor pens (favored in Mexico and Guatemala), the air temperature, water temperature, and humidity are quite high, and the photoperiod is controlled by timers. In the tropics, the frogs may breed continually throughout the year and generally are allowed to mate naturally, but sometimes breeding is induced with hormone injections. Strict hygiene is

American Bullfrog   *The only frog widely bred for the culinary trade, the American bullfrog is now farmed in several tropical countries, especially in Southeast Asia and Central America, to supply the enormous global demand for frog's legs. However, commercial production currently supplies only about 15 percent of the market, (Europe alone imports 6,000 tons of frog's legs annually), and wild frogs, caught mainly in India, still provide the balance.*

Photo: Bruce MacQueen, Shutterstock.com

practiced as the frogs, like all skin-breathers, are extremely susceptible to bacterial and fungal infections.

Breeding fecundity depends upon a bullfrog's age. Females laying for the first time may produce only 1,000 eggs, whereas farmed females three years old have laid up to 20,000 eggs. The eggs are removed to separate tanks indoors, and hatch after 48 hours, but the larvae do not swim until they have absorbed the rest of their yolk sac when they are three days old, and are then moved to rearing tanks. They are herbivorous at first, eating mainly algae, but become carnivorous when they metamorphose into froglets (by growing legs and absorbing their tails); this occurs between 45 and 90 days, depending on the temperature and their precociality. They are then trained to eat flakes (usually containing 40 percent protein), on which they grow rapidly, some being ready for market three months after metamorphosing. They are anaesthetized by freezing, and are then slaughtered and processed in certified facilities. Their hind legs are cleaned in chlorinated water and bagged; and, when frozen at 5°F (-15°C), have a shelf life of six months.

4 Reptiles

Despite the long history of reptile keeping, which goes back to ancient Egypt, reptile breeding until quite recently was an irregular and usually accidental occurrence. Consequently, unlike certain fish (the goldfish and common carp) that have been domesticated for several thousand years, captive reptiles were probably not bred regularly until the 1860s, when soft-shelled turtles (*Pelodiscus sinensis*) were farmed in Japan. By 1907 their production had risen to 60,000 animals annually, so they were possibly the first examples of reptiles beginning the process of domestication. Diamondback terrapins (*Malaclemys terrapin*) were apparently "raised" in the United States for the restaurant trade early in the last century, but it is unclear if they were captive-bred or were raised from eggs collected in the wild—which, of course, does not produce self-sustaining populations nor contribute to domestication. Turtle farming in Louisiana did not begin until just after World War II, and Taiwan began farming turtles in the 1950s. But it was not until well into the second half of the last century that the breeding of other reptiles, in zoos and related establishments, became more than just isolated incidents.

The incredible advances in reptile husbandry and breeding in the last three or four decades has resulted in the establishment of many captive colonies to supply the great demand. The boom was fuelled in part by the increased interest in reptilian pets, with an estimated nine million now kept in the United States alone (of which a large percentage are captive-bred), plus similar interest in Europe and Southeast Asia. It has been estimated that 3 percent of households in the United States have pet reptiles. The breeding boom was also encouraged by the inability of many wild populations to supply those pets or the growing local and global demand for reptile meat, skins, and shells, due to their severe depletion or government restrictions. Regrettably, there is still a very large trade in wild-caught reptiles for the general pet trade; many arrive in poor health and are then purchased by people who are unfamiliar with their basic requirements.

Three factors are involved in the captive production of reptiles—commerce, zoos, and research laboratories. Breeding for commercial use is by far the largest, while zoos and research establishments have relatively minor roles. Commerce includes the turtle and crocodile farmers, as well as the breeders of iguanas, many other lizards, and numerous snakes. As with any form of commerce, production is based on demand, and for several species the trade is terminal. The production of the crocodile trade is totally terminal, with farmed specimens marketed at a precise weight and size; and turtles in commerce are destined as food, the ingredients of traditional medicines, or live animals for the pet trade. Snake breeders produce live animals for the pet trade and hobbyists. Snake farming in Southeast Asia is not yet a major industry, but undoubtedly will soon be providing snakes for the restaurant trade as the supply of wild snakes dwindles. Pet keepers are consumers, as pets are normally single animals, with no opportunity to reproduce. For at least one lizard—the common iguana—commerce involves its large-scale production for local consumption and exportation as pets, whereas all other captive-raised lizards in commerce are produced for the pet trade. Hobbyists, or amateur herpetologists, obviously do not enter this category. They are unlikely to consider their animals as pets, and some even specialize in cobras and pit vipers, which are hardly pet species. The regular breeding of cold-blooded animals and their eventual domestication differs from that of the mammals and birds as their young are in most cases abandoned at birth or as eggs. Parental care is rare and therefore cannot assist the young to adapt to artificial conditions.

In keeping with the progress made in bird and mammal breeding, reptile reproduction in zoos flourished in the last decades of the twentieth century. Husbandry improved, especially through the careful attention to their animal's environmental requirements, and keepers began to specialize. However, zoos and related institutions—collections of wild animals where conservation and research have important roles in addition to display—have a different purpose to the commercial producers. Their mandate is to maintain pure stock, characteristic of their wild ancestors, in which they are assisted by species survival plans, studbooks, cooperative breeding programs, and collaboration with governments in the animal's countries of origin, often establishing breeding facilities there. Consequently, many zoos are involved in the captive breeding of some of the world's most highly endangered reptiles, several from just a few founder individuals, which escalates the process of domestication. Reptiles have also been involved for many years in science classrooms and research laboratories, but these have generally been wild-caught specimens, and in fact many still are. However, numerous self-supporting captive colonies have also been maintained for many years, including a large leopard gecko colony, and thousands of pit vipers on a pharmaceutical company's snake farm.

The regular breeding of wild animals often raises concerns about purity. Fortunately, most commercial breeders of crocodiles and turtles maintain the purity of their stock, which still resemble their recent ancestors. In the pet trade, some loss of purity has occurred in snakes through hybridization, but the most visible changes have resulted from the production of mutants. Hundreds of color and pattern morphs, some of them quite startling, have been bred in many species of snakes and lizards. Natural morphs occur in all wild populations and appear sooner or later in captive-bred animals, when they are perpetuated and increased by selective breeding. For many species, breeding is no longer simply a possibility, as it was a few decades ago. Now, for leopard geckos, bearded lizards, Burmese pythons, corn snakes, and many others, it is simply a case of pairing a mutant to a mutant and awaiting the almost inevitable highly unusual offspring. The process of domestication is obviously well underway for these reptiles. As they have been controlled and bred for many generations, the process is unavoidable, even though it may be unintentional. Simply retaining the best breeders or the tamest specimens, coupled with unnatural diets, artificial mate selection, and compromising their territorial and social behavior habits, affect animals' behavior and eventually their physiology and anatomy, but not necessarily their purity.

More than any other group of animals, the captive breeding of reptiles, especially the large-scale production of "new" lizards and snakes, has introduced a whole new aspect and vocabulary to pet keeping, as genetics, breeding strategies, mutant production, and color pigmentation are now involved. The basic breeding of wild animals, which for years involved keeping a pair or group together, hoping they would breed, and then letting them do so for as long as they could, is a thing of the past. Now, the breeding of reptiles involves inbreeding, linebreeding, and artificial selection. Inbreeding is the mating of closely related animals, the degree of inbreeding depending upon the closeness of the pair, which may be mother and son, or father and daughter, or siblings. Genetic diversity is lost and there is increased expression of recessive traits, and the consequent mutation or permanent change in a cell's DNA sequence produces a mutant individual, differing from the wild or normal type of the species. Linebreeding is the mating of animals of the same bloodline, a milder form of inbreeding involving uncles, aunts, and cousins. Selective breeding or artificial selection, is the intentional mating of animals chosen by humans to achieve or eliminate a specific trait. Breeders choose animal's mates for them based entirely on their desirable traits so that the offspring inherit these characteristics. Genetically, the dozens of leopard gecko and bearded lizard color and pattern morphs now available are usually recessive traits, which is actually a genetic disorder that appears only in animals that have received two copies of a mutant gene, one from each parent. This obviously requires more than one animal in the breeding population to carry the gene, and for two of these animals to mate, which rarely happens in the wild; but it is encouraged in captivity by inbreeding, when offspring are purposely mated back to one of their parents. Breeding a mutant to a mutant then accelerates the rate of variation. A recessive trait is one that is invisible when combined with a dominant trait, in the gecko's case the typical cream-colored, black-spotted individual. For it to become visible, a gecko must have two copies of the allele for the trait, one from each parent.

The desired, but sometimes unexpected, outcome of this virtual frenzy of mutant-production is a great range of colors; and the names of the various pigmentations, and even their absence, are now also part of the vocabulary of advanced reptile keeping. Black pigment or melanin is responsible for the normal wild leopard gecko's dark eyes and many spots. If a mutant's pigment is excessive, the condition is known as hypermelanism or just melanism. In contrast, hypomelanism is a reduction of the pigment, and amelanism (or albinism) is when the black pigment is totally lacking. It is the same with red pigmentation; hypererythrism is excessive red pigment, anerythrism is when the red pigment is lacking, and so on. The animal is said to be hypermelanistic or amelanistic, hypererythristic or anerythristic.

The pigment-containing cells, or chromatophores, responsible for reptile's skin and eye color, are grouped according to their color. Melanophores synthesize black or brown pigments. Xanthophores produce yellow pigmentation, and an individual in which they are completely lacking is considered axanthic and appears much darker than normal, sometimes almost melanistic. Iridophores are reflective platelets that produce the iridescence of certain snakes, in the well-named rainbow boa, for example. An animal lacking iridophores is said to be aniridic, but such mutants are rare. The absence of xanthophore and melanophore genes produces white individuals, and to avoid confusing them with albinos or white specimens they have been called "snows." Reptile breeders involved in the production of unique mutants have a distinct advantage over the breeders of other animals, because of the presence of these pigments and the ability to combine the genes of several mutants to produce really striking individuals.

The result of all these mutations, of pattern as well as color, and the size differences now also appearing, are all highly visible signs of domestication. In addition, a less obvious but sure sign of man's complete control of wild animals, is their acceptance of totally artificial diets, not just replacement ones that resemble their natural foods. It also indicates that the demand is large enough to justify their commercial formulation, production, and marketing. Like the feeding of chickens or laboratory rodents, pelleted foods are now available for a wide range of reptiles, especially lizards. There are proprietary diets for carnivorous lizards such as tegus and monitors, special diets for the omnivorous blue-tongued skink and the bearded dragon; and flavor-coated dried flies for the insectivorous geckos and anoles, To enhance palatability, additives called "odor enhancers" scent the artificial diets with the smell of a mouse or a cricket. Artificial diets for the herbivorous lizards, such as iguanas, contain beta carotene—the natural yellow pigment that is the precursor for vitamin A—instead of the vitamin itself, which is potentially toxic to them.

■ TURTLES

The name turtle is widely used in North America for all the members of the order *Testudinata*. In Europe, only the aquatic species are called turtles, the terrestrial ones being known as tortoises and the semiaquatic ones terrapins. The aquatic turtles are the most plentiful reptiles in commerce, produced in their millions for food or as pets. Consequently, they are the most domesticated cold-blooded land vertebrates, and were likely the first reptiles to be bred regularly and in quantity—in Japan in the middle of the nineteenth century. The modern trade in captive turtles involves mainly the farming of freshwater species, and the few marine turtles produced are far outnumbered by the wild ones that die annually in fishing nets or are caught illegally. The commercial trade in turtles is largely terminal, whether they are produced for their products or as pets, when they are normally

Soft-shelled Turtle *The very aggressive and carnivorous soft-shelled turtle's snake-like neck and elongated snout allow it to breathe while its body is submerged or buried in the mud. Two species are farmed in large numbers in Southeast Asia, mostly for the Chinese market, where they are honored dishes at banquets.*
Photo: Photos.com

kept singly. The captive-raising of terrestrial tortoises commercially forms a small portion of the overall turtle trade, and breeding for conservation is generally on even a much smaller scale, except for the very successful Galapagos tortoise breeding program.

The current international demand for freshwater turtles is staggering, and involves mainly three regions of the world. The major market is Southeast Asia, primarily for food and shells—an important ingredient in traditional Chinese medicine—and their preserved eggs. In Europe and North America, the trade involves mainly pet turtles. Two species dominate this trade. In North America, especially in Louisiana, the main species is the red-eared slider (*Trachemys scripta elegans*), while in Southeast Asia the favorite turtle for farming is the Chinese softshell (*Pelodiscus sinensis*).

Turtle farming has been practiced in the Orient for many years, perhaps even before the soft-shell *Trionyx sinensis japonensis* was raised in Japan over a century ago. However, more recently the bulk of Asian turtle farming has been mainly to supply the Chinese market. Taiwan began farming soft-shelled turtles in the 1950s, and Malaysia and Thailand in the 1980s. In 1996, six million farmed turtles were exported from Thailand to China, and in 1997 1.5 million captive-bred turtles were exported from Taiwan. In 1998 Indonesia exported 370,000 turtles, and in the first nine months of 1999 Malaysia exported almost 2.5 million turtles, of which 1.5 million were farm-raised. But consumption in China continued to soar, in large part due to the fact that soft-shelled turtles are honored dishes at Chinese banquets, and the new economy made them more affordable. The increased demand resulted in large importations of wild turtles from other Asian countries, especially Vietnam, Laos, Myanmar, and Bangladesh, further depleting their wild turtles and endangering species already threatened. This situation stimulated turtle farming throughout Southeast Asia—to supply the demand, not to conserve turtles.

Reported production figures for Asian farmed turtles vary, but most estimates indicate that farms outside China produce about 10,000 tons of turtles annually, a figure unfortunately still equaled by the number of wild-caught turtles, and the consensus is that the wild populations cannot withstand such annual harvesting. In Thailand and Malaysia there are numerous turtle farms with holdings of 25,000 turtles each, in total producing several million hatchlings annually. There are apparently hundreds of small family-run turtle farms in China, and the limited reports of large-scale commercial production include a farm in Hainan province that has 50,000 breeding turtles. The favorite species in all the countries is the Chinese soft-shell (*Pelodiscus sinensis*); the wattle-necked soft-shelled turtle (*Palea steindachneri*) is also farmed. Hard-shelled turtles farmed in China include Reeves turtle (*Chinemys reevesi*), with one million hatchlings produced annually, and the Chinese three-striped box turtle (*Cuora trifasciata*), whose shell is believed to have cancer-curing properties. Also raised in China are the Malayan box turtle (*C. amboinensis*), the Chinese box turtle (*C. flavomargiata*), and the yellow pond turtle (*Mauremys mutica*). Several hundred thousand hatchlings of the Chinese stripe-necked turtle (*Ocadia sinensis*) were produced in Taiwan annually in the 1990s. To meet the increasing demand, even many of the red-eared slider hatchlings imported into Southeast Asia from the United States are now also raised for farming. Despite the demand, it is now reported that many turtle farmers are changing over to the more lucrative business of farming tilapia, gouramis, and prawns.

The production of captive turtles obviously reduces the pressure on wild populations, although wild-caught turtles still augment even the captive-raised breeding stock in some regions. Unfortunately, turtle farming only reached such proportions due to the dwindling of wild stocks in the first place. One major criticism of Chinese turtle farming concerns the considerable hybridizing that occurs, both accidentally and intentionally. Unusual specimens that have recently appeared in the pet trade, and have been described as new species, are believed to be hybrids.

Turtle farms in the United States continue to be the world's major producers of hard-shelled turtles, primarily the red-eared slider (*Trachemys scripta elegans*), which is the most commonly kept reptile in the world. Turtle breeders in the United States also raise several other species, including the common snapping turtle (*Chelydra serpentina*), the river cooter (*Pseudemys concinna*), map turtle (*Graptemys geographica*), musk turtle (*Stenotherus odoratus*), mud turtle (*Kinosternum subrubrum*), and the southern painted turtle (*Chyrsemys picta dorsalis*). Few turtle mutants have appeared, and the red-eared slider is believed to be the first to produce color variants that have been perpetuated. The most common is the albino, now frequently seen in the pet trade, in which the hatchlings are bright yellow. A pastel morph of the slider, which is pale yellow with pale green and red markings, is also bred, and albino common snapping turtles and musk turtles have also been offered for sale.

The commercial farming of the other aquatic turtles, the marine species, is centered on the green turtle (*Chelonia mydas*), which has proved to be the easiest to maintain and breed. The Grand Cayman Turtle Farm is the longest established sea turtle farm, begun in 1968 with eggs collected from the wild. But from 1978 it has relied on eggs laid by its own turtles—both wild-caught and captive-hatched and raised animals. In addition to the green turtles, some hawksbill turtles (*Eretmochelys imbricata*) have been produced there. Controversy over the definition of farmed and ranched turtles, especially regarding their licensing for trade, has been partially solved by CITES defining a farm as an establishment that maintains animals that breed and so maintain self-supporting populations. A ranch is defined as a place that hatches eggs collected from the wild, and the hatchlings may be used for commercial purposes or released. Consequently, only the Grand Cayman operation warrants the title of farm, whereas the turtle "breeding farms" that exist in Surinam, Indonesia, Australia, Thailand, India, and Reunion are more correctly ranches, as they rely on the collection of eggs laid by wild turtles, although their purpose may be to return the hatchlings to the sea. This practice, termed "headstarting," lowers the risk of predation, which is very high during the marine turtle's natural hatching and early life at sea, and is now frequently employed to benefit turtles of several species.

Many species of land turtles or tortoises are also bred commercially, for sale as pets or to reptile hobbyists. Generally, their production is not called farming because it is on a much smaller scale than the production of aquatic species. Captive breeding includes the South American rain forest species—the red-footed tortoise (*Geochelone carbonaria*) and the yellow-footed tortoise (*G. denticulata*). Also, African species such as the pancake tortoise (*Malacochercus tornieri*), the leopard tortoise (*Geochelone pardalis*), and the African spurred tortoise (*Geochelone sulcata*). Also several "Mediterranean tortoises"—the Greek tortoise (*Testudo graeca*), marginated tortoise (*T. marginata*) and Hermann's tortoise (*T. hermanni*). In view of the increasing rarity and value of the star tortoise (*Geochelone elegans*), a very attractive species from India, and of Madagascar's radiated tortoise (*Geochelone radiata*), their captive breeding, even if for commerce, is welcome. An aspect of turtle development of considerable interest to breeders who artificially incubate their turtle's eggs, at least to those who are producing future breeding stock, is the influence of temperature on the sex of the offspring, as the developing embryos do not have sex chromosomes so their sex cannot be determined genetically. Their gonads are sensitive to temperature (thermosensitive), and in the pond turtles and box turtles (*Emydidae*), for example, incubation temperatures above 78.8°F (26°C) produce mostly female hatchlings; and at lower temperatures most of the babies are males.

In addition to commercial turtle production, considerable breeding success has also been achieved by zoos and related organizations. A classic example is Perth Zoo's breeding program for the western swamp turtle (*Pseudemydura umbrina*), the world's rarest species, which was virtually extinct in the wild. Captive breeding began in 1988, and the zoo has produced 50 hatchlings annually for release. The other great success occurred on the Galapagos Islands, where the breeding program for giant tortoises at the Charles Darwin Research Station has raised 1,000 Espanola giant tortoises (*Geochelone elephantopus hoodensis*). Another very successful institution, with both turtles and crocodilians, is the Madras Crocodile Bank, India's premier wildlife establishment, which breeds several native turtles including the star tortoise, the Travancore tortoise (*Indotestudo travancoria*), the Indian soft-shell (*Aspideretes gangeticus*) and the red-crowned roof turtle (*Kachuga kachuga*).

■ SOME OF THE SPECIES

Red-eared Slider (*Trachemys scripta elegans*)

The red-eared slider is a common river and pond turtle of the Mississippi River Basin, occurring naturally from Virginia to Iowa and south to Georgia and Texas. But it has been widely transported, and populations elsewhere in North America, such as in California and Maryland, are believed due to human agency. Released or escaped turtles have also colonized Japan, Australia, and several European countries. A greenish-yellow turtle about 10 inches (25 cm) long, it is usually recognizable by the broad red stripe behind its eye, at least in the males, but the female's stripes may be indistinct. Occasionally the stripe is completely lacking in both sexes, and in some individuals it is yellow. The red-eared slider prefers still waters, muddy bottoms and dense water weed, and at the northern end of its natural range it must hibernate and spends October to April buried in the mud. It is a very wary animal, and for quick access to the water it basks on partially submerged logs and on mats of floating vegetation rather than on the pond banks.

Sliders breed throughout the summer months, beginning in early June, and usually lay two or three clutches, with up to 15 eggs per clutch. They may travel some distance from their pond to a suitable nest site, where they dig a shallow hole and bury the eggs. Their incubation period is 60–75 days, and the last clutch may hatch so close to winter that the young stay in the nest, protected by antifreeze compounds that prevent their intracellular fluids from freezing. They are omnivores, their natural diet comprising a mix of pond weed and animal life including snails and dead fish. Captive animals are given a high protein supplement that may include trout chow, waterproof (floating) fish food, or dog chow, but the bulk of their diet is vegetable matter such as dandelion leaves, green leafy vegetables, carrot tops, shredded raw carrots, and apples.

The slider is the most popular pet turtle in North America and the most domesticated of all the turtles, bred in large numbers for over half a century, and turtle farms in the United States continue to be the world's major producers. Turtle farming began in Louisiana just after World War II, and by the late 1960s the annual production was said to be 15 million animals. However, in 1975 concerns over the risk of salmonella infection caused the Federal Food and Drug Administration to ban the sale of turtles less than 4 inches (10 cm) long in the United States, a trade which at the time amounted to about nine million animals annually. Overseas markets were then expanded, especially in Europe and Southeast Asia, and the salmonella problem was eventually overcome by eliminating the bacteria on the eggs through sterilization. Millions of laboratory-tested and certified disease-free baby turtles are produced annually, accounting for over 90 percent of the 10 million reptiles exported from the United States each year.

Soft-shelled Turtle (*Pelodiscus sinensis*)

This unusual turtle is a native of China, Korea, Vietnam, Taiwan and Japan. Its shell or carapace is soft, leathery, and quite flexible, as it lacks the horny plates of the other turtles and terrapins. It is also dorso-laterally compressed (pancake-like) and is about 10 inches (25 cm) long when adult. Soft-shells have long snake-like necks and an elongated snout. They are the quickest of all turtles, their partially-webbed hind feet propelling them very fast in the water; and despite being highly aquatic, they can also move quickly on land. They are totally carnivorous, and in the wild they eat fish, crustaceans, mollusks, and aquatic invertebrates, snapping their necks out to capture prey in their powerful jaws. Their long necks and elongated snouts allow them to breathe while their bodies are submerged, and they often lie buried in the mud in shallow water with just their snouts above the surface. They can remain completely submerged for several hours, extracting oxygen dissolved in the water through gill-like vascular papillae in their throats, as they draw it into their pharynx and then expel it. They lay up to 20 eggs twice annually, burying them in a hole dug into a sandy riverbank.

Despite their need for an animal-protein diet, soft-shells are the most commonly farmed turtle in Southeast Asia, with many large farms in Thailand, Vietnam, and China. One farm on the island province of Hainan, begun in 1983 with animals collected from the wild, is said to hold 30,000 breeding turtles in large ponds, thick with water weed and guarded by ferocious dogs. A farm in Vietnam has 80,000 breeding-age soft-shelled turtles, sells 200,000 annually as future breeding stock, and each day sends 440 pounds (200 kg) of live turtles to Hanoi's markets. Farmed animals have escaped in Thailand and are now established there, and specimens imported into the United States as pets have escaped or been released and are now thriving in Florida and Hawaii.

Galapagos Tortoise (*Geochelone elephantopus*)

This is the world's largest tortoise, and consequently the largest land reptile, which reaches a length of 4½ feet (1.3 m) and weighs up to 650 pounds (295 kg). It continues growing until it is 30 years old. These giant reptiles are restricted to the islands of the Galapagos Archipelago, about 600 miles (965 km) off the coast of Ecuador; there they have suffered terribly, directly and indirectly, from human intervention for the past four centuries. The first major threat to their survival was their harvesting by visiting sailors, whalers, and pirates, beginning in the seventeenth century, which is estimated to have reduced their limited population by 200,000 tortoises. More recently the threat has been indirect, from the introduced rats and feral cats and dogs that eat the eggs and hatchlings, and there are currently only about 10,000 tortoises left.

The giant tortoises show considerable variation on their islands as they evolved to take advantage of the different environments. The wet and lush ground-level vegetation on the highlands of Santa Cruz and Isabela has resulted in tortoises with dome-backed carapaces, short legs, and short necks. On the drier islands of Pinta and Espanola, the tortoises have "saddlebacks," longer legs and an upturned front to their carapace, allowing them to reach up for higher vegetation. Consequently, there were 13 subspecies until recently, but only 11 survive, and some of those are very rare. Their natural distribution is five subspecies on the island of Isabela, and one each on the islands of Pinzon, Santiago, Espanola, San Cristobal, Santa Cruz, and Pinta—but only a single male of that island's subspecies—*G. e. abingdoni*, known as "lonesome George"—survives.

Since the establishment of the Charles Darwin Research Station (CDRS), several subspecies have been assisted by the practice known as headstarting. This involves collecting the eggs laid by wild tortoises, hatching them at the Station, and raising them to a size when they are safe from rats and dogs before releasing them. This does not contribute to domestication of course, but many giant tortoises are being raised in a manner that does. They were first bred outside the islands in 1954 by the Honolulu Zoo, which has since hatched many and distributed them to other zoos. In 1969 the San Diego Zoo joined in partnership with the CDRS; the zoo now has one of the largest captive colonies outside the islands, and has raised many hatchlings. The greatest success, however, has been achieved with the Espanola tortoise (*Geochelone e. hoodensis*), through captive breeding at the CDRS. In 1963 the last 14 giant tortoises surviving on the island were transferred to the station's breeding center at Puerto Ayora on Santa Cruz. The first captive-raised tortoises were returned to the island of their ancestors in 1975, and in 2000 the 1,000th tortoise raised at the breeding center was returned to Espanola, where there are currently 800 tortoises, all captive-raised animals.

■ LIZARDS

The regular production of lizards and their subsequent domestication is mainly commercially inspired to supply the pet trade, and only one species—the common green iguana (*Iguana iguana*), is a dual-purpose animal, farmed in several Central America countries as a local source of food and for the overseas pet trade. Consequently, the most commonly imported reptiles into the United States in recent years were captive-raised iguanas, with imports averaging one million annually, and large numbers were also shipped to Europe and Japan. Farming lizards is therefore not as widespread, in numbers or localities, as the farming of freshwater turtles.

The availability and cheapness of imported baby iguanas discouraged serious efforts to breed the species in North America and Europe, and the increasing number of lizard breeders in recent years concentrated on other easily bred species, especially two that produce mutants readily. These are the leopard gecko (*Eublepharis macularis*) and the bearded lizard or bearded dragon (*Pogona vitticeps*), of which dozens of color and pattern mutants are now available. The leopard gecko is one of the most popular lizards kept as pets and certainly the most frequently inbred, linebred, and selectively bred species, whose domestication is not in doubt. A female may lay 20 eggs in a breeding season, during spring and summer, in clutches of two eggs at a time. Leopard gecko mutants for color and pattern began appearing in 1991, and size variation is also involved, with some mutants larger than the normal wild gecko. In keeping with its name, the wild gecko is creamy-white spotted with black. Recessive traits have produced albinos, blizzards and patternless geckos. Linebreeding over many generations has produced tangerine, super hypo-tangerine, carrot tail, jungle and hypo-melanistic. The super hypo-tangerine is a very attractive lizard, with a tangerine colored body, and tail with a bright orange base and white end, spotted with black.

Other breeders concentrate on chameleons, now increasingly popular despite their more difficult diet, as they need live foods rather than the artificial pelleted foods now acceptable to many lizards. The veiled chameleon (*Chamaeleo calyptratus*), the panther chameleon (*Furcifer pardalis*), and Jackson's chameleon (*Chamaeleo jacksoni*) are now often bred. Blue-tongued skinks (*Tiliqua scincoides*), tegus (*Tupinambis teguexin*)—available in several color morphs—and numerous species of monitor lizards (*Varanidae*) are also regularly produced. The most interesting lizards available on the commercial market as captive-raised hatchlings are the beaded lizard (*Heloderma horridum*) and Gila monster (*H. suspectum*)—the world's only venomous species.

Jackson's Chameleon   *Despite their strict insectivorous nature, chameleons are increasingly kept as pets and several species are now breeding regularly. Jackson's chameleon is one of these, a spectacular and popular large species that has been more commonly available than "exotic" species due to its introduction and establishment in the Hawaiian Islands.*

Photo: Michael Ledray, Shutterstock.com

The zoo breeding of lizards has kept pace with successes in all other departments, and its growth can clearly be seen in the annual captive-breeding records published in the International Zoo Yearbooks. In addition to their individual efforts, many zoos are involved in breeding consortia for specific endangered lizards, both in their own facilities and in the species' country of origin. While the pet trade was flooded with green iguanas by the million, zoos were applying their limited resources to other iguanas, especially the rock iguanas of the genus *Cyclura,* with collaborative efforts to save them. These lizards are endemic to islands in the Caribbean, and are highly endangered through loss of habitat and hunting. The Indianapolis Zoo has been involved since 1995 in the captive breeding of the Grand Cayman Island blue iguana (*Cyclura lewisi*)—probably the world's rarest lizard, with just 50 surviving in the wild—and has produced young annually. The Jamaican iguana (*Cyclura collie*), reduced to perhaps 100 individuals by mongooses on its native island, is now also the subject of breeding programs involving a consortium of U.S. zoos, with the first offspring ever hatched outside Jamaica born at the Indianapolis Zoo in 2006.

A spectacular earlier zoo success involved the komodo dragon (*Varanus komodoensis*), the world's heaviest lizard and possibly the longest (which may be Salvadori's water monitor), which was first bred outside its native Indonesia at Washington's National Zoo in 1992. The zoo has since raised 80 hatchlings that have been distributed to 30 zoos, where many have also since reproduced. An even earlier zoo success involved the Round Island Gecko (*Phelsuma gueuntheri*). Round Island lies just north of Mauritius, a once-beautiful tropical island that was virtually denuded of its vegetation by introduced herbivores. With several of its endemic species in danger of extinction, the late Gerald Durrell collected 16 geckos in 1976 and took them to his Jersey Wildlife Park, where over 300 were raised and distributed to other zoos.

Lizards have also proved their value in biological studies, and several self-supporting colonies have been maintained for long enough to be considered virtually domesticated. The leopard gecko was a favorite laboratory species long before it became the most commonly bred captive lizard in North America. A self-sustaining colony of over 200 leopard geckos was maintained at London's St. Thomas's Hospital for many years for dermatological research. Beginning in the early 1960s, their success showed how easy it was to breed geckos in small plastic containers, certainly one of the first instances of closed colony lizard breeding. At Virginia Tech's Department of Fisheries and Wildlife Sciences, a large laboratory colony of western fence lizards (*Sceloporus occidentalis*) is maintained for controlled toxicological studies.

■ SOME OF THE SPECIES
Chameleons (*Chamaeleonidae*)

Chameleons are the world's most unusual lizards, and one of the most recognizable. They are medium-sized laterally-compressed animals, best known for their extended eyes, their ability to change color, and their very long tongues. Their heads are roughly triangular in shape and may have horns, a casque, or a crest. With one known exception, they are entirely insectivorous. Chameleons live in Africa, southern Europe, southwestern Asia, and on Madagascar and other islands in the Indian Ocean. They are diurnal and are ideally adapted for arboreal life, as their tails are prehensile, they have opposable digits that give a good grip on branches, and are slow and purposeful climbers. Spectacularly colored, chameleons have pigmented skin and change color for a number of reasons, mostly connected with emotions, temperature, and condition. During courtship, females change color to warn off a male, who then changes color to impress her. Color changes occur when animals are receptive, after breeding, when sick, too hot or too cold, or stressed for any other reason, but they do not change color for cryptic purposes to match their surroundings. These changes are made possible by the expansion or constriction of the chameleon's three layers of pigmented skin cells, called chromatophores, that are covered by the epidermis—the outer transparent skin, which is the one they shed when moulting. The chameleon's eyelids have fused and have evolved into extended cone-shaped eyes that can rotate and be focused independently. The familiar rocking motion they make upon spotting prey has range-finding value. Their long extensible tongues have a hollow and sticky pad at the end, and can be extended for just over double the animal's body length in adults and almost four times the length of babies.

Wild-caught chameleons are still being imported, but their breeding is fortunately becoming more commonplace, despite their specialized diets—they are strictly insectivorous and must have live food. However, after traditionally being considered very difficult animals to maintain for any length of time, let alone breed, several species now breed regularly in North America and Europe and captive-raised babies are frequently available, some in a number of color morphs. The most commonly bred species is the veiled chameleon (*Chamaeleo calyptratus*) from the Arabian Peninsula, followed by the panther chameleon (*Furcifer pardalis*) and Jackson's chameleon (*Chamaeleo jacksoni*). The latter is a large and impressive species that reaches a length of 14 inches (36 cm). Males have three large horns—one on the snout called the rostral horn, and two over the eyes called the ocular horns—which they use in territorial battles. Depending upon the subspecies, females may have only one horn or none at all. This spectacular lizard is a native of East Africa, especially Kenya, but its exportation from there was banned in 1981. However, one of its races, *C. j. xantholophus,* is now quite common in the pet trade as a result of being introduced almost 30 years ago onto the Hawaiian Islands, where it is now well established on Oahu and Maui. In addition to exports from Hawaii, it is frequently bred, and captive-raised specimens are regularly available. It is a live-bearer, whereas most chameleons lay eggs.

Bearded Dragon (*Pogona vitticeps*)

From the very arid environment of central Australia, the bearded dragon has become one of the most popular pet lizards. It is an agamid lizard, a member of the widespread family *Agamidae,* which contains about 350 species. There are several species in Australia, but it is the only one to have become a commonly kept pet. Dragons are spectacular animals, with a large, broad, triangular-shaped head and a throat pouch or beard that is inflated during courtship or when in threat display. They have spiny extensions around the head, on their throats, and along the sides of the body, but they are not as hard and sharp as they appear. When adult their flattened bodies measure 22 inches (55 cm) long, including their tail. They can run upright on their hind legs for short distances. Wild bearded dragons are usually pale reddish-brown, but there is considerable variation and they can darken their color. They are omnivores, and their captive diet comprises pink mice, mealworms, and crickets, plus finely chopped fresh greens and shredded apple, carrot, and zucchini. Some breeders have changed from this "natural" diet to a completely artificial pelleted food.

Dragons are highly personable animals and become very tame. They love to sunbathe, although in their natural habitat they hide in burrows from the extreme midday heat. They breed freely, can be kept in pairs, and, if provided with a sand box with at least 12 inches (30 cm) depth of slightly moist coarse sand, they will lay up to 30 eggs, which are generally removed and incubated in moist vermiculite. They have responded to selective breeding to achieve or eliminate a specific trait, and numerous mutants are now available. They include several red dragons, such as sandfire red, which is the most popular morph, and blood dragons; plus hypo-pastel, tangerine, snows, and citrus. The leucistic dragon is white with dark eyes (so it is not an albino), its color resulting from the inability of its chromatophores to produce pigment. True albinos were recently also produced.

Common Iguana (*Iguana iguana*)

This large lizard has been a favored food item in Central and South America for many years. In Central America the wild populations were able to sustain the hunting until midway through the last century, when forest was cleared for cattle ranching. They also faced increasing hunting pressure, from both humans and their traditional predators—large constrictors, crocodiles, and birds of prey—while their eggs and hatchlings are vulnerable to rats and dogs. Despite its alternate name of green iguana, it is quite variable in color. Juveniles are usually vivid green and may be banded, but they darken with age, males often assuming a shade of orange and females olive; but some have pronounced bluish and turquoise hues. However, despite the color variations, no subspecies are currently recognized. Escaped or released iguanas are now living ferally in Florida, Texas, and Hawaii. When mature at the age of four years, males may be 6 feet (1.8 m) long, of which over half is tail, and the females are about two-thirds the size. Their prominent characteristics are a very long banded tail that is used as a whip in self-defense, an impressive crest along the spine from the nape to the base of the tail, and a massive dewlap.

Iguanas live near water in lowland forest and are highly arboreal, climbing with their long and sharp claws. They are also excellent swimmers and drop into the water when threatened. They are the only farmed lizards, their husbandry pioneered by Dr. Dagmar Werner in Central America in the 1980s, as an operation to protect the dwindling populations of wild iguanas traditionally harvested for food. It developed into a massive "cottage" industry, providing not only food and animals for restocking depleted areas, but also many hatchlings for the overseas

Common Iguana *A multipurpose lizard, the iguana has been favored as a source of food in its native habitat since man arrived there long ago, but more recently it has become a popular house pet, with over one million hatchlings exported annually to North America and Europe. To meet this large demand, and to continue to provide food in regions where forest clearance has reduced their habitat, iguanas are now farmed in several Central American countries.*
Photo: Clive Roots

pet trade. The iguana pens are made of sheet metal sunk into the ground to form a smooth unclimable surface. A bamboo and palm thatch shelter is provided for them, plus a sand-filled nesting chamber entered via an artificial tunnel. Iguanas can breed when they are two years old, and dig a hole in their sand box in which they lay up to 30 eggs, which incubate for three to four months. The captive iguanas are fed fruit, leaves, and iguana chow, a homemade product consisting of boiled rice, ground meat, and fish and bone meal. Although iguana farming originally began to conserve them for local consumption, the demands of the pet trade encouraged their greatly increased production; in recent years at least one million hatchlings have been exported annually to the United States, and many have also been shipped to Europe and Japan.

■ SNAKES

Throughout history snakes have played a major, although contradictory, role in human culture and folklore. Symbols of evil in some religions, they are worshipped by others. Eaten by some cultures, they are usually at the bottom of western zoo animal popularity surveys. But until today in the Western world, they have never had such a close relationship with man. Except for dedicated herpetologists, snakes, especially the large constrictors, were considered the prerogative of zoological gardens just three decades ago. Now they are increasingly kept as house pets, and imports into Europe and the United States from the tropics in recent years have numbered in the tens of thousands annually. Commercial breeders are now also supplying the pet trade and hobbyists with an unprecedented selection of mutants of many species. Their regular captive breeding, already to many generations, and the complex selective breeding involved in the production of new variants on a frequent basis, leaves only one conclusion. As unlikely as it may seem, many species of snakes must now be considered domesticated. They may only be in the early stages of the process, but it is certainly underway, and seems likely to continue. The booming interest and demand, the improved knowledge of their husbandry, and the many accessories now available to improve their housing, health, diet, and regular reproduction are indications that it will.

The constrictors are one of the most successful groups of captive snakes, with many species now reproducing regularly. They include numerous pythons such as the reticulated, Burmese, carpet, ball, green tree, and Children's pythons; and several boas, including the common, emerald, rainbow, and Madagascar boas. Captive-bred North American snakes include rat snakes, king snakes, garter snakes, corn snakes, and fox snakes, and European species such as the grass snake, four-lined snake, and Aesculapian snake. The increase in captive breeding and the demand for new mutants has led to the intentional production and perpetuation of many color and pattern morphs, but fortunately not much hybridization. The mutants, morphs, or aberrations that appear occasionally in the wild have been eagerly acquired by breeders. They are usually recessive, and to be perpetuated the offspring must inherit a pair of the same genes from each parent, thus requiring more than one animal in the breeding population to carry the gene. Consequently this rarely happens in the wild, but is encouraged in captivity by inbreeding, when offspring are purposely mated back to one of their parents. The offspring of a normal snake and a mutated one will have inherited both normal and mutated genes, and are said to be heterozygous. While their phenotype (their outward appearance) is observable, their genotype (their genetic identity) does not show externally, and can only be determined through breeding.

The ball python (*Python regius*) is probably the most popular python in the pet trade, and for many years has been exported from West Africa, especially Togo, Benin, and Ghana, in shipments amounting to tens of thousands annually. These were wild-caught snakes despite claims of "farming" them, and mortality was high initially, but they are favored because of their docility and ease of handling. Fortunately, in the hands of experienced herpetologists, the ball python is now one of the most regularly bred snakes, and dozens of mutants have been produced by selective breeding. So named for their habit of curling into a tight ball, even wild ball pythons show a great range of color and pattern and rarely are two snakes alike, but basically they have a dark background with golden-brown blotches. The most popular mutants are albinos, which first appeared about 15 years ago and result from a single recessive gene that causes amelanism (lacking dark pigment); the snake's normal dark background is consequently white, and its blotches are yellow. Several other variations of the albino have since been produced, including one in which the white background is tinged with lavender. Interbreeding the basic mutants, of which there are about 40, has produced many "designer morphs" such as super pastel jungle, pastel caramel albino, and silver bullet. Ghost ball pythons, a simple recessive color form, are hypomelanistic—having reduced melanin—resulting in a pale "ghostly" color.

Other pythons now being bred in large numbers include several subspecies of the carpet python (*Morelia spilotes*), especially the coastal or Queensland carpet python (*M.s. mcdowelli*), one of the smaller species; and Children's python (*Liasis childreni*), another small species from northwestern Australia. The green tree python (*Chondropython viridis*) now reproduces readily, in several mutants as well as its normal bright green with white markings. This is one of the snakes in which the hatchlings differ completely from their parents, being brown, yellow, or maroon, and change to their adult coloring when they are about eight months old. The captive-produced morphs include blues, melanistic mutants that are black spotted with green, and calicos, which have a yellow background color speckled with tiny green and orange spots.

Despite their size and potential danger, even the largest constrictors, the reticulated python (*Python reticulatus*) and the Burmese python (*Python molurus bivittatus*), are now kept and frequently bred. The reticulated python's record length is about 33 feet (10 m), and captive animals regularly grow to 20 feet (6 m) in length and can weigh 200 pounds (90 kg). The Burmese python is usually slightly smaller, although one does currently hold the world record for the largest captive snake. Wild-caught mutants of these giant snakes have occasionally been displayed in zoos, but captive-bred morphs began to appear about 15 years ago and now include albinos, calicos, tigers, super tigers, jaguars, and yellow-headed reticulated pythons, all of these names reflecting their color or pattern.

The boa constrictor (*Boa constrictor*) is another common snake imported for the pet trade in large numbers; these imports include several subspecies and numerous geographic color variations. It is now also breeding regularly, and selection for a particular characteristic has resulted in many pattern and color morphs. Albinos have been available for many years, the lack of melanin resulting in a white-and-yellow snake with a contrasting red-and-orange tail. Recent mutants include melanistic snakes that are virtually all black but have a golden belly, and the salmon, a pink snake with dark red saddles. Another South American boa, the emerald tree boa (*Corallus caninus*), a very similar snake to the green tree python, is also being bred, although its husbandry is more challenging. There is wide variation in their basic color and the extent of their markings in wild populations, and to take advantage of these, a much wider range of color and pattern variations must eventually be produced.

Exotic species are not the only snakes to benefit from increased captive breeding, and the most commonly kept and selectively bred species, with dozens of recent mutants, is actually a North American native—the corn snake or red rat snake (*Elaphe g. guttata*). The classic example of modern snake domestication, it was one of the first species to experience serious selective breeding, beginning in the early 1970s. It is now available in many color variations, including albinos, creamsicles, snows, and blood-reds. Pattern mutants include striped, zigzag, and motley—a simple recessive mutant that results in a series of irregular blotches or stripes. The corn snake has been hybridized with several subspecies of rat snakes (*Elaphe obsolete*) and king snakes (*Lampropeltis getula*).

Snake farms or snake parks, where venom is collected regularly from native species for the production of antivenin and other biomedical uses, exist in several countries. The Butantan Institute at Sao Paulo is the most famous, established in 1901, and where the first monovalent (containing one kind of antibody) and polyvalent (containing several antibodies) antivenins were produced. It now has a collection of 50,000 venomous snakes, the world's largest, and is a leading producer of immunobiologicals and biopharmaceuticals. The Queen Saovabha Snake Park in Bangkok collects venom from the many Southeast Asian snakes kept there. Just north of Sydney, many tiger snakes are "milked" regularly at the Australian Reptile Park, which for 30 years has been the country's largest supplier of snake and spider venom for antivenin production. Breeding may occur in these collections, but it is not the primary goal. However, in the world's largest snake monoculture—a facility housing just a single species—breeding essential to maintain a regular supply of snakes for the extraction of venom takes place. In 1981 the Swiss pharmaceutical company Pentapharm founded a self-sustaining snake farm at Uberlandia, in southwestern Brazil. It currently houses 10,000 Brazilian lanceheads (*Bothrops moojeni*), a highly venomous species from which the individual compounds in the extracted venom are used in the research and production of blood clotting agents.

■ SOME OF THE SPECIES

Burmese Python (*Python molurus bivittatus*)

Pythons are constrictors like the boas and are usually all included in one family, the *Boidae*. One of the major differences between the two, and the most obvious, is associated with reproduction—pythons lay eggs and boas give birth to live young. Pythons have quite large external spurs—claws on either side of the vent that aid the mating process—and some have heat sensing labial (lip) scales to help them locate their prey. The Burmese python, a subspecies of the Indian python, is a native of Burma, eastward throughout Southeast Asia, and Indonesia. It is the most popular large constrictor in the pet trade, and import records show that almost 30,000 hatchlings were imported in the United States annually in the last years of the twentieth century and in this century. Unfortunately, they tend to outgrow their owner's interest and may then be released, and National Park Service employees have captured many large specimens in recent years in Everglades National Park.

Captive Burmese pythons grow very large. The current record for the heaviest captive snake is held by a Burmese python aged 21 years, living at the Serpent Safari Park in Gurnee, Illinois. In 2005 "baby" weighed 403 pounds (182 kg) and was 27 feet (8.2 m) long. The constrictors have lethal potential long before they reach such an enormous size, and 10-foot-long (3 m) specimens have killed human infants and even a 15-year-old boy. Some city bylaws prohibit the keeping of constrictors over a specific length, and large individuals require a very secure, escape-proof enclosure. Despite the obvious risk, Burmese pythons are often kept as pets, and captive-born snakes are generally calm and friendly. Breeders are now producing a multitude of mutants, of which the most common one is the albino, a white snake with pale gold markings. The pattern morph called the granite is particularly unusual, the snakes having a speckled, granite-like appearance with a pale head and red eyes. A very recent addition to the range of color and pattern is

Burmese Python Hatchlings   *Despite their potential for very large size, Burmese pythons are popular snakes in the pet trade and have been imported in large numbers to meet the demand. They breed readily, however, and captive-hatched babies are now regularly available and far more suitable as pets, being healthier and usually more tractable than wild-caught snakes.*
Photo: Photos.com

the albino granite, a bright orange snake covered with small, pale, irregular markings. The rarer mutants are in great demand and may be worth several thousand dollars, so they are potentially a good investment for experienced breeders.

Corn Snake (*Elaphe guttata guttata*)

The corn snake, or red rat snake, is the most colorful of the rat snakes, with considerable variation in color but generally with reddish blotches bordered with black on a pale yellowish background. It is one of the smaller species, averaging 39 inches (1 m) long, although individuals up to 70 inches (1.8 m) have been recorded. It has a wide distribution across the southern United States and into Mexico, where it prefers drier habitat such as dry river bottoms, rocky hillsides, and pine barrens. Like the other rat snakes, it is mainly terrestrial but is a good climber. The corn snake is the most popular pet snake and the most frequently bred captive snake in the world. It has responded to improved breeding techniques, and there has been extensive production of attractive mutants since the first albino was bred in 1962.

Breeders in North America annually produce thousands of mutant corn snakes, many of them resulting from single recessive genetic mutations. Hypomelanistic snakes are those in which the black pigment is reduced, heightening the red, white, and orange colors, which appear very vivid. In amelanistic individuals, also called red albinos, the melanin pigment is absent, so they are yellow and white with orange blotches or "saddle" markings, or they may be solid orange. They have red eyes, and their colors are produced by reflective platelets or iridophores in their skins. When their red color is affected by a genetic aberration, they are anerythristic when it is absent and hypererythristic when there is far more red pigment than normal, when they are called red corn snakes. Snow corns, which are double recessive mutants resulting from a combination of the anerythristic gene and the amelanistic gene, lack both the red and black pigment and are white with pink markings. Pattern morphs have also been produced—striped ones, in which the normal blotches are replaced with five longitudinal lines; and zigzags, in which the blotches are connected to form a zigzag pattern. Corn snakes have also been hybridized with various subspecies of the king snake, including the California king (*Lampropeltis getulus californiae*) and variable king snake (*L. mexicana thayeri*), and, when crossed with Emory's rat snake (*Elaphe guttata emoryi*), have produced "creamsicle" albinos.

Brazilian Lancehead (*Bothrops moojeni*)

The genus *Bothrops* is a group of highly venomous pit vipers called lanceheads, occurring in Central America, South America, and some Caribbean islands. A familiar species is the fer-de-lance (*Bothrops atrox*), the most feared snake in Central America, northern South America, and Trinidad. Pit vipers are the most specialized of all nocturnal snakes, adapted for hunting in total darkness. With their sophisticated heat-sensing systems, in the form of a facial "pit" containing thousands of nerve endings sensitive to the heat radiating from their prey, they can locate warm-blooded prey in complete darkness.

The Brazilian lancehead is a snake of the central Brazilian plateau or cerrado, a seasonally dry region. It shows considerable variation in color, from pale-brown to reddish-brown and gray, with a general pattern of triangular-shaped markings, often with dark edges and separated by a paler color, and has the typical triangular head of the pit viper. A mature snake is usually 4–5 feet (1.2–1.5 m) long. The lancehead prefers the gallery forests—along river banks and the moist adjoining grassland, where its main prey are lizards and frogs. It has extremely toxic venom, containing enzymes that degrade the proteins in the blood plasma, causing hypotension (loss of pressure) and circulatory shock.

Although zoos and snake parks breed venomous snakes occasionally, the Brazilian lancehead is the only species of venomous snake that can truly be considered in the early stages of domestication. It well illustrates the fact that modern or recent domestication is not necessarily synonymous with docility and harmlessness. The Swiss pharmaceutical company Pentapharm has maintained a breeding facility for the lancehead at Uberlandia in Minas Gerais since 1981, to ensure a regular supply of snake venom used in research and the production of a range of pharmaceuticals in the fields of anticoagulants, antifibrinolytics, and haemostatics. The Uberlandia facility currently has a breeding population of 10,000 lanceheads, housed in low-walled outdoor pens, each providing an environment resembling the snake's natural habitat. This is an enormous operation, the largest snake-breeding monoculture. Even the world's major zoological gardens keep only about 500 specimens of reptiles of all kinds, not just snakes.

■ CROCODILES

The recent trade in the three other groups of reptiles—the turtles, lizards, and snakes—has thrived on the supply of pets, but captive crocodilian commerce is now based almost entirely upon farming for their hides and meat. The trade in crocodiles as pets was once quite widespread, especially with the exportation of thousands of baby spectacled caiman from South America annually; fortunately, it has declined considerably. This is mainly due to international controls, the realization that their eventual size makes their care as pets difficult and potentially dangerous, and perhaps because so many captive-bred snakes and lizards are currently available. The unregulated hunting of crocodiles for their hides, coupled with the loss of habitat and the subsequent decline of most species, resulted in the introduction of controls by CITES, with all crocodilians now included in either Appendix I, in which trade is only permitted under exceptional circumstances, or in Appendix II, in which trade is strictly regulated and monitored. The continued demand for crocodile products, however, led to an increase in farming, which took the pressure off the wild populations and is considered the savior of several crocodilian species.

Crocodile farms originally started with eggs collected from the wild, as they still are in some regions, but many are now self-supporting. According to CITES, collecting eggs to produce crocodiles is "ranching," and not "farming," in which the crocodiles are actually breeding and the operations are therefore self-sustaining. Ranching obviously does not contribute to domestication, as the crocodiles are not being captive-bred. However, when eggs are collected from the wild and hatched and raised for commercial purposes, there is usually a requirement to return some of the hatchlings. Crocodile farmers generally keep species native to their region, and CITES has registered many farms for specific crocodilians. In Thailand there are several farms registered to produce the Siamese crocodile (*Crocodylus siamensis*). In Cuba, the native Cuban crocodile (*C. rhombifer*) is now farmed; in Mexico, Morelet's crocodile (*C. moreletii*) is kept; and in Honduras, the American crocodile (*C. acutus*) is farmed. The giant saltwater crocodile (*C. porosus*) is widely kept in Australia, New Guinea, Indonesia, Malaysia, Thailand, the Philippines, and Singapore, all countries within the natural range of the species. The equally large Nile crocodile (*C. niloticus*), the obvious species for farming in Africa, is produced in South Africa, Zimbabwe, Botswana, Kenya, and Tanzania, and CITES has also registered farms for this species on the islands of Madagascar and Mauritius. In Florida, the American alligator (*Alligator mississippiensis*) is widely farmed. Crocodile farmers obviously cannot specialize in local species if none exist, such as in Israel, where American alligators have been farmed since 1981 at Hamat Gader, near the Sea of Galilee.

The American alligator suffered severely from unregulated hunting, which was banned in 1962, although poaching continued until 1967 when it was included on the first endangered species list. Alligator farming was approved in the United States in the 1980s, and there are now 30 farms in Florida. A multimillion-dollar industry, it provides 15,000 hides and a lot of meat annually. Australia's native species, the saltwater crocodile and the freshwater or Johnstone's crocodile (*Crocodylus johnstoni*), were seriously depleted during many years of unregulated hunting. A quarter of a million of each species were shot in northern Australia during the late 1950s and throughout the 1960s. So few remained that hunting became uneconomical, and the practice was stopped, first by the government of Western Australia in 1969, followed a few years later by the Northern Territory and Queensland. The crocodiles recovered, and from the late 1970s licenses were issued for farming.

Both species are bred in Australia on 17 farms, which hold a total of 50,000 specimens. The same species is ranched in New Guinea, where farms were started with eggs collected from well-populated rivers such as the Sepik and the Fly.

With its widely scattered wild distribution, and management programs in many areas, the current population of the Nile crocodile is estimated at between 250,000 and 500,000 individuals, so it is not an endangered species. For the past two decades the UNFAO has assisted the development of crocodile production in several developing nations. They began in Zimbabwe and have since been involved in other countries in south and east Africa, although these operations are actually ranching, not farming, as eggs are collected from the wild, hatched, and raised. This arrangement does have conservation value, however, as in Zimbabwe the ranchers are required to return a proportion of their hatchlings to Lake Kariba, where the eggs were collected. So the people benefit from the sale of hides, and the species also benefits. The high mortality rate of hatchling crocodiles makes egg hatching and the repatriation of youngsters a good conservation tool.

Thailand is the center of crocodile farming in Southeast Asia, with many CITES-registered farms including the Samut Prakan Crocodile Farm, just outside Bangkok, which was founded in 1950 and is therefore probably the longest established one in the country. Many filial generations of crocodiles have since been produced. The collection, consisting of about 30,000 crocodiles of several species, but mainly the Siamese crocodile, is the world's second largest, after the Darwin Crocodile Farm. Hybridization of the saltwater crocodile and the Siamese crocodile occurs there, and elsewhere in Thailand, as the hybrids are valued for their superior leather and high tolerance of their often poor and crowded captive conditions.

Crocodiles are carnivores and require a diet of animal protein for many years, as they are slow growers. The farming of meat-eaters is uneconomical if it costs more in protein consumption than the product returns, so farming crocodiles is therefore unrealistic in regions where the diet of the local populace is lacking in protein. Crocodiles are fed by the cheapest possible means, and their diet usually includes old battery hens past their egg-laying usefulness, slaughterhouse offal, and fish heads.

In addition to the many crocodile farms and ranches, which both in their own way contribute to the conservation of crocodilians, several breeding operations function purely for conservation. One of the most important of these facilities is the Madras Crocodile Bank, which has 2,400 crocodiles of 14 species and has bred 5,000 mugger crocodiles (*Crocodylus palustris*) for release since its inception in 1976. It has also been breeding gharials (*Gavialis gangeticus*) since 1989. Unlike their wild relatives, the muggers at the Bank have been producing two clutches of eggs annually, which is unique in crocodiles and may be due to their better diet.

■ SOME OF THE SPECIES

Saltwater Crocodile (*Crocodylus porosus*)

The largest crocodile, and the world's largest reptile, the saltwater crocodile is also called the estuarine crocodile. It has reached 23 feet (7 m) in length and a weight of one ton (1,017 kg), although larger specimens have been claimed but not substantiated. A marine animal, and thus very saltwater tolerant, it has evolved lighter scales as an adaptation for its highly aquatic lifestyle. Also, like the tube-nosed seabirds that can excrete excess salt and therefore drink seawater without harm, the saltwater crocodile has many glands (called osmoregulators) in the mucus membrane of its tongue that secrete sodium chloride and control the internal concentration of salts. It can therefore live in water with up to 60 percent salinity, and its habitat is the coasts of southern India and Sri Lanka, eastwards in an arc encompassing Bangladesh and Myanmar, Malaysia, Indonesia, the Philippines, and northern Australia. It can spend days at sea and is able to cross large expanses of open ocean—for example, between the Indian coast and the Andaman Islands, a distance of 690 miles (1,100 km). In northern Australia it lives along the coastline and river estuaries and in mangrove swamps.

The saltwater crocodile is a very aggressive and dangerous reptile, a notorious man-eater responsible for many fatal attacks on humans, and which probably accounts for more human deaths than all species except the infamous Nile crocodile. Its natural foods include axis deer, sambhar deer, monkeys, and wild boar in India, and kangaroos and feral water buffalo in northern Australia. Throughout its range it preys on fish, turtles, sharks, water birds, and snakes. Its nest is a mound of vegetation and soil close to the water's edge, or on a mat of floating vegetation, where the heat of decomposition from the decaying leaves produces warmth for the egg's development. The mound also raises the eggs above the water level or water table, as immersion and oxygen deprivation for more than a few hours is fatal to the embryos. A female lays about 50 eggs, and their incubation period is 80 to 90 days depending on the environmental temperature. During this time she guards the nest, and opens it up when the hatchlings call. When necessary she helps them hatch by breaking open the shells with her mouth, or by pressing them with her legs or body. She very carefully carries the hatchlings to the water, but they may occasionally be injured in the process. The young form a creche and she stays nearby, communicating with them and guarding against predators, and they rush back to her when alarmed and may take shelter in her throat pouch.

The saltwater crocodile is the most farmed crocodilian. It is kept in New Guinea, Indonesia, Malaysia, Thailand, the Philippines, and Singapore; but most farms are in Australia, where the captive crocodile population (which includes some freshwater crocodiles) is said to number about 50,000. The largest farm is the Darwin Crocodile Farm, which houses about 36,000 animals and marketed 8,000 in 2006, a figure that is expected to rise to 15,000 per year in 2010. They are harvested for their meat and hides when about three years old and are about 6 feet (1.8 m) long. Australian crocodile farming does not totally contribute to domestication, however, as only the farms in Queensland and Western Australia rely primarily on captive breeding to supply their breeding stock. Farmers in Northern Australia are currently allowed to harvest eggs laid by wild crocodiles.

Saltwater Crocodiles *The largest reptile, reaching a length of 23 feet (7 m), the notorious and very aggressive saltwater crocodile is a frequent man-killer. Contrarily, it is also the most farmed crocodilian, with many farms in Southeast Asia and Australia. The market is mainly for their soft belly hide, which is used to make quality leather goods, and they are "harvested" when they are about 6 feet (1.8 m) long.*
Photo: Grishin Konstantin, Shutterstock.com

Chinese Alligator (*Alligator sinensis*)

This alligator, one of only two species in the world (the American alligator being the other), has been the subject of a captive breeding program for a quarter of a century. It is one of the rarest crocodilians, reduced to less than 1,000 animals when the Chinese government established the program in 1980. At the time it had vanished from most of its historic range, the victim of population growth, loss of habitat, and indiscriminate capture, and occurred only in the lower reaches of both the Yellow and Yangtze rivers and a few watercourses between the two. Protection in the wild was not enforced, and villagers also killed them when they entered their rice paddies. Crocodilians have poor temperature tolerance, and the Chinese alligator, whose range extends north to about 31° latitude, just a little lower than the American alligator's range, becomes dormant in riverbank burrows from early November to the end of March.

To protect the remaining few animals, the Anhui Research Centre for Chinese Alligator Reproduction (ARCCAR), established in 1980, stocked several farms with 200 alligators captured in the wild, plus the hatchlings from several hundred wild-laid eggs. Local people were also paid to bring in captured alligators and eggs, but initially the eggs were unhatchable as they had been turned—probably just in handling—which tears the membranes, unlike incubating bird's eggs that must be turned frequently to prevent the yolk sticking to one side. The first captive-born young were produced in 1988, and the farms were registered by CITES in 1992. In 1993 the management of the farms was leased to a Thai company, and other breeding centers have since been opened. The ARCCAR's current population numbers about 10,000 captive-bred animals in 26 small protected areas, and over 1,000 alligators are raised annually. There are currently believed to be about 200 alligators left in the wild in China, but some captive-bred animals fitted with VHF transmitters have already been reintroduced.

5 Birds

Prior to the last two centuries, birds were kept mainly to provide food for the growing human population. The domestication of the chicken, goose, and duck several thousand years ago all resulted from the demand for more meat and eggs; and economics also influenced the more recent control by man of the guineafowl, pheasant, and turkey. New domesticates appeared during the nineteenth century when pet-keeping became popular, especially with the availability of several colorful birds from Australia. Then in the last century, dozens of new species began to breed regularly. The prehistoric bird domesticates were social species and were all vegetarians (seed-eaters or grazers), whose diets were easy to reproduce, even in medieval times, although they generally just foraged for themselves, in summer at least. The new pet species of the nineteenth century were also seed-eaters and social species, giving rise to the belief that domestication was the prerogative of gregarious, non-carnivorous birds. The regular reproduction of such solitary breeders as emus and cranes, and the breeding and hybridization of totally carnivorous species as hawks and falcons in the last century, has shown that this assumption must be revised.

Until about four decades ago, parrots were considered difficult to breed, mainly because they were not easy to sex, and "pairs" were kept for years hoping they would reproduce. The ability to sex birds invasively through laparoscopy initially solved that problem, but now it is even easier and less stressful on the birds through testing the DNA of their feathers or blood. Falcons were also very difficult to breed, due to inadequate housing rather than incorrect sexing, as most are sexually dimorphic, but given the isolation they need, they now reproduce readily, and are even hybridized to produce better birds for falconry. Many other bird groups have also shown a tremendous increase in their willingness to reproduce under man's control, and the subsequent availability of captive-bred birds for future breeding, rather than wild-caught ones, also improved reproductive success.

Two very familiar pet birds, the budgerigar and the cockatiel, arrived in England from Australia in 1840 and 1845, respectively; and the zebra finch, another popular Australian cage bird, was imported into Europe even before the budgerigar, but the exact date is not recorded. Most of their selective breeding and hybridizing, and the production of color mutants, did not occur until the twentieth century, however. The first mutant cockatiels did not appear until after World War II. Other parakeets, including the red-rumped parakeet and several species of rosellas, were also exported from Australia in the middle of the nineteenth century and bred readily in European aviaries. The lovely Gouldian finch was discovered in central Australia by artist John Gould only in 1841, but captive birds were soon producing mutants. Lovebirds have also been bred since the mid-nineteenth century, and many mutants are now available. Some of these birds are now so domesticated that societies have developed show standards for them. The Zebra Finch Society has exhibition standards, for type, markings, and color, that have little relationship to wild birds but are required to effectively judge the numerous mutants. Their changes to date are mostly concerned with color and pattern, except for the budgerigar, of which the British show bird is considerably larger than its ancestors. Although at this stage these new birds are obviously not as fully domesticated as the chicken, turkey, or duck, with their great range of mutations, especially for size, there can be no doubt that the process is well under way. They are the bird kingdom's equivalent of the recently domesticated golden hamster, chinchilla, and gerbil, and their domestication is ongoing.

Commercial diets, scientifically formulated to provide a bird's complete nutritional requirements and usually in pellet form, are now available for most birds, including waterfowl, parrots, and game birds, which signifies a market large enough to warrant production. They are typical of the diets produced for long-term domesticates, such as cats, dogs, chickens, and laboratory rats, but some breeders do not use them. While they may be adequate for semi-free-ranging waterfowl and pheasants able to scratch and graze in their aviaries, such boring repetitious diets are considered psychologically unsuitable for some birds, especially parrots, even though they may be nutritionally complete.

Three factions are involved in the care of the many birds that are now being domesticated. They are the pet keepers, the commercial breeders, and the aviculturists. Pet keepers are consumers, not producers, as they generally keep single birds, but they are derived from stock that has been completely controlled for many generations, and each year their breeding becomes more simplified. So there are really just two producers, and their goals differ. The commercial breeders are primarily concerned with agricultural and economic incentives—production efficiency, "breed" improvement, and new mutant development—and some of their production is terminal. In contrast, the objectives of most private aviculturists, plus those in zoos and related institutions, are to maintain pure birds that remain characteristic of their wild counterparts in morphology and behavior—although in the confines of captivity the former is indeed difficult and the latter virtually impossible.

■ THE BIRD KEEPERS

Pet Keepers

A pet is defined as a tame animal kept for amusement or companionship, generally in the house. With the increasing difficulty of importing wild-caught birds due to conservation and health regulations, the bulk of the species now kept as house pets are produced by commercial breeders. Tame, usually hand-raised, and accustomed to people, they are far more suitable anyway than wild-caught birds. Although finches, doves, and a few small softbills may be kept indoors in flight cages, the term "companion animal" really only suits the parrots (a collective name for cockatoos, macaws, parrots, and similar psittacine birds) that can be allowed out of their cage and handled regularly. Their almost free-breeding habits of the past few decades has allowed a level of pet-keeping that the wild populations could never support. A similar situation occurred in the nineteenth century, when the exportation of thousands of wild cockatiels and budgerigars from newly settled Australia could not meet the European demand and encouraged their large-scale breeding and domestication. It occurred again when Australia banned the exportation of its native birds in 1959, which resulted in increased captive breeding of parakeets and grassfinches.

Pet owners probably form the largest group of exotic animal keepers, but little breeding occurs as they generally keep individual birds, their contribution to the domestication process being more indirect—providing a market for the breeders. Favored birds include hanging parrots, caiques, the green Amazon and gray African parrots, and cockatoos and macaws of many species. There are also domestically produced small seed-eaters such as zebra doves, diamond doves, cutthroat finches and long-tailed grassfinches; and a few of the smaller softbills, especially Pekin robins, shamas, and to a lesser extent hill mynahs.

The loss of breeding potential and genetic material has always been a major criticism of keeping single animals, but this stems mostly from the time when trapping wild birds was a drain on natural populations, when there were high losses during shipment, and the birds did not breed. This no longer applies where breeding is concerned, as the onus is on the commercial breeders, and frequent breeding is in their interest. But a great deal of interest also lies in the production of mutants, which results in the loss of genetic material. The domestication of bird species bred for the pet trade will occur at a much faster rate than those maintained by many aviculturists, zoos, and related institutions, because breeders are producing birds for their suitability as house pets. They are consequently selecting for tameness and acceptance of the captive environment, and are artificially incubating the eggs, hand-raising the chicks, and practicing hybridization and inbreeding.

Aviculturists

Aviculture is usually defined as the keeping and breeding of wild birds. Individuals are involved, plus institutions such as zoos, bird gardens, and specialist collections. The practice of aviculture is also important in government facilities, which produce endangered native species for reintroduction. Aviculturists benefit from their career or avocation in various ways, including the interest and challenge involved in the captive management and breeding of birds; the opportunity to assist conservation through captive breeding, study, and possibly reintroduction; and the accumulation of considerable data on bird behavior, nutrition, and reproduction. As a result of their efforts, the birds also benefit in most of these areas. Aviculturists' interests may lie in the husbandry of a specific group such as softbilled birds, pheasants, waterfowl, or parrots, or they may specialize in cockatoos, touracos, grouse, or pigeons. They generally maintain the purity of their birds, but may produce mutants. The size of their holdings varies, from one or a few pairs of a species

Blue Crowned Pigeon   *One of three species of crowned pigeons from New Guinea, the blue crowned breeds frequently in zoos and private collections. The largest and most beautiful of the pigeons, they have a slow reproductive rate, laying only two eggs per clutch, but they compensate for this by breeding until they are 25 years old.*
Photo: Clive Roots

of particular interest, to large mixed collections, and to some spectacular ones that rival and occasionally even surpass those of the major zoos in species and numbers. These hobbyists should not be confused with pet keepers. They are serious bird keepers who do not consider their birds as pets.

The value of aviculture lies not only in the production of offspring characteristic of the species, but also in the recording of their experiences. A vast amount of data has been collected and published over the years that is of value to other breeders and to biologists, especially those attempting to protect endangered populations in their native lands—"in situ" conservation, as this is now called. It is generally accepted that data recorded on captive animals can make significant contributions to scientific knowledge. Aviculture also has great conservation value, especially in the production of pure specimens for eventual re-introduction. However, if aviculturists are breeding birds regularly, even out-crossing with stock acquired from other breeders, they cannot avoid domesticating them.

These hobbyists are a new phenomenon in numbers only, for in the nineteenth century and well into the twentieth, wealthy connoisseurs sent their own collectors overseas for live specimens, and many of their acquisitions were new to science. Live trapping is no longer necessary for many species of birds, and in fact for most is impossible due to export and import bans. Many are usually readily available anyway from other breeders, so in general aviculturists are now producers and not consumers of wildlife. In fact some groups, of which the pheasant breeders are a shining example, have maintained several species for years from very limited initial stocks, on occasion just a single pair, and the opportunity to outcross the descendants of these founders with wild imports was only recently possible.

Although breeding has always been an important aspect of aviculture, as with the pet bird trade it became more essential in recent years when legislation reduced the availability of wild-caught specimens, beginning with Australia's ban on the export of its native birds. This provided incentive to increase breeding efforts, but cannot detract from the fact that aviculture has been an essential component of bird protection programs for many years. With the reduction of many birds in the wild due to the exploitation of their environment, overpopulation, loss of habitat, climate change, and the introduction of aliens, the importance of captive breeding has long been appreciated and has been endorsed by both the World Conservation Union (formerly the International Union for the Conservation of Nature) and the World Wildlife Fund.

There are many examples of captive breeding by individuals and institutions saving endangered species and successfully reintroducing birds. Early this century, the wood duck was believed to be following the passenger pigeon into extinction in the wild, and the captive breeding program that produced over 2,500 birds for release is credited with forming the nucleii of breeding colonies throughout eastern North America. More recently, captive breeding has produced cheer pheasants for release in Pakistan, eagle owls for return to Germany's forests, and many peregrine falcons in North America and barn owls in England to bolster wild populations. It has saved the trumpeter swan, the whooping crane, the rails of Lord Howe Island and Guam, the Bali mynah, the Micronesian kingfisher, and the Californian condor from almost certain extinction. The national breeding programs conducted by several countries, by New Zealand at Mt. Bruce and at Patuxent in the United States, have shown the value of captive breeding for conservation and reintroduction.

Although aviculture and pet bird keeping, or the "bird fancy," were originally quite distinct because of their involvement with wild and domesticated birds respectively, there is now considerable overlapping between them, which can only increase. Some aviculturists are so successful in their production of pure specimens that commercialism may be involved in their disposal. Also, several species, such as the greenfinch, ring-necked parakeet, and quaker parakeet—all originally within the aviculturist's domain—have now been controlled for many generations, selectively bred to create and perpetuate color mutants and hybrids, and are now obviously domesticated birds. The process of domestication has already begun for many species of parrots, finches, pheasants, raptors, and waterfowl, which have been breeding regularly for many generations. The continuing improvements in captive reproduction and the reduced availability of wild birds will eventually result in the domestication of many other species, requiring a redefinition of the term "aviculture."

Several of the world's rarest waterfowl have been saved from extinction by captive breeding, from very small initial numbers. The total world population of the Hawaiian goose or nene stems from 25 birds, and the reintroduced "wild" population on the islands is continually reinforced with releases of captive-bred geese. The Laysan teal (*Anas platyrhynchos laysanensis*), a long-isolated subspecies of the mallard, was virtually exterminated on Laysan Island when introduced rabbits denuded the vegetation. Although discovered only in 1891, by 1912 there were only seven birds left. Captive propagation saved the teal and some have already been returned to their island, now cleared of rabbits. Both species of flightless teal living on sub-Antarctic islands south of New Zealand have been saved by captive breeding. The Auckland Island flightless teal (*Anas a. aucklanidca*) and the Campbell Island flightless teal (*Anas nesiotis*) are both being bred at New Zealand's National Wildlife Centre.

However, even though efforts are made to keep birds pure and characteristic of their species, captivity still produces unforeseen changes, as shown by grouse. The predominantly vegetarian species have a very long intestine and cecum to digest the highly fibrous pine needles and heather that form the bulk of their wild diet. Captive red grouse (*Lagopus lagopus scoticus*), fed a diet of concentrated commercial chicken pellets, experienced shortening of the gut due to the lack of fiber. Their ceca, where fiber is digested by bacterial fermentation, were reduced in length by one quarter, and their small intestines were only half their natural length, a potential problem if they were destined for reintroduction.

Commercial Breeders

The commercial production of birds now plays a major role in the exotic animal industry, providing live birds, mostly for the pet trade, and game birds and ratites for their meat. With extensive, and often quite intensive, breeding facilities, the production of pet birds has considerable relevance to conservation, as these birds—parrots, cockatoos, and macaws, for example—were formerly wild-trapped, so it has reduced exports from the countries that still allow them. These breeders also, but to a lesser extent, supply specimens to the hobbyist, such as the falconer, the waterfowl aviculturist, or the softbilled bird enthusiast. Many of these regularly bred commercial species are therefore in the early stages of domestication, and eventually will become as domesticated as the budgerigar. Tame birds, habituated to humans, are preferred, so captive-bred parrots are usually hand-raised. However, although natural "wild type" birds may be raised, for many species the trade involves breeding practices that emphasize hybridization, inbreeding, and the production and perpetuation of color mutants. Some extremely attractive ones are now being produced, and their initial high commercial value places them in the forefront of breeding efforts. Poor parents or not, all that is really required of them is to lay fertile eggs, because incubators and dedicated hand-raisers do the rest.

The family *Psittacidae* (the parrots and their relatives) provides all the birds that can be considered companion animals in the home. To man they are the most intriguing of all birds anyway and have always been preferred as house pets. They have been popular since the days of the ancient Romans and Greeks, and early seafarers brought them back to Europe from the tropics. The pet trade in parrots for many years comprised the small domesticated forms such as budgerigars, cockatiels, and lovebirds, and a great many wild-caught birds. During the twentieth century, there was a large international trade in imported birds as pets, especially of cockatoos, macaws, Amazon parrots, and African gray parrots. This situation has changed dramatically, and after so many years of being considered a difficult group of birds many species are now reproducing very successfully, and hand-raised birds are regularly available. In recent years there has been a great deal of intensive research on the psittacines, involving their aviary design, nutrition, reproduction, and medical care. For those whose interests lie in aviculture and the breeding of pure forms to aid conservation, in genetics and the development of color mutants, or in the production of birds for the pet trade, parrots have responded. They are adaptable, intelligent, gregarious, and dexterous, and many are excellent mimics. They are the only birds that can truly be called companion animals. In the last quarter of the twentieth century, great strides were made in parrot breeding, and many species proved as adaptable as the budgerigar and cockatiel.

The other group of birds always considered difficult to breed were the birds of prey or raptors, especially the diurnal species. Prior to the middle of the last century, peregrine falcons and goshawks were even considered vermin whose capture and trading was legal, and their breeding was a rare occurrence. Zoos, despite recording longevities of over 30 years for several species, had a poor record of breeding the diurnal hawks and a mediocre one for the nocturnal owls. Lack of interest, birds kept in solitary for life, wrongly sexed pairs, and accommodation lacking the seclusion needed for breeding all contributed to the poor results. Reproductive success was not achieved until the late 1960s, when the drastic decline of falcons from the effects of organochlorine pesticides led to increased interest in raptors and their plight. Captive colonies of falcons were established, and the first species to benefit from the research into the effects of pesticides was the American kestrel, the continent's smallest raptor. The production of birds for falconry also escalated, fuelled in part by the sudden ease with which they were breeding, plus the increase in protective legislation that made wild falcon acquisition difficult. Between 1970 and 1975, the number of captive-bred falcons doubled, with half of those produced by private falconers; within a few years, breeders in the United States were raising 500 falcons annually. Like the breeding of parrots, it is no longer a major feat to propagate birds of prey, and almost 100 forms (species and subspecies) have reproduced in captivity, including 23 kinds of falcons. Thousands of peregrine falcons and kestrels have been raised, plus hundreds of other species such as lanner falcons, saker falcons, lagger falcons, prairie falcons, and gyr falcons. Although they have not received the attention given to the falcons, and certainly not from the institutional breeders, several species of hawks—such as the northern goshawk, Harris's hawk, and the European sparrow hawk—have also proved just as responsive to improved captive husbandry.

There are now many commercial facilities producing captive-bred raptors for falconers. Their breeding has been enhanced by artificial insemination, artificial incubation, and hand-raising. Manipulation of the photoperiod has induced kestrels to breed out of season—in the winter between consecutive spring breeding seasons. For artificial insemination, semen is collected regularly during the breeding season by the manipulative method used in poultry, and through the use of imprinted males that willingly copulated with a collecting device and even mated with their keeper's hat. This has increased fertility by 15 percent over natural matings in kestrels, and they are now bred almost as easily as chickens; but there is currently more interest in producing the more valuable falcons. The prairie falcon is another easily bred species, initially favored by American falconers but now bypassed in favor of peregrines and gyr falcons, and especially by hybrid falcons. Domestication of the falcons, as predicted by falconer Frank Beebe in 1964, has indeed arrived.

Game birds are also raised commercially, and in large numbers. The industry is mainly terminal, as most are raised for their meat; but there is also the aspect of the sale of breeding stock, and some are kept for egg production. Although it was kept as a pet and a singing bird in Japan in the twelfth century, the Japanese quail (*Coturnix japonica*) has more recently been farmed commercially for the restaurant trade. In the nineteenth century it was selectively bred for egg and meat production and spread to China, Korea and Hong Kong between 1910 and 1940, during Japan's empire expansion. The stocks were largely lost during World War II, but after the war the survivors were mated with birds from China. Currently France is a leading producer, with a national flock of 6½ million birds in 2004. The Japanese quail is also valuable in laboratory research in the United States, where it has been used since the 1950s in physiological, embryological, nutritional, and genetic research. The guineafowl is another original domesticate, kept by the ancient Romans but since redomesticated, that has also seen a great revival in interest in recent years as a bird for the table, rivalling the chicken in the tenderness of its flesh. France is also the leading producer of guineafowl, with 10 million birds producing 42,000 metric tons of dressed carcasses in 2004.

Canada is a major producer of game birds (ring-necked pheasants and quail), with the province of Quebec being the main supplier, raising almost nine million birds annually with a retail value of $24 million. Selective breeding is practiced to improve growth rates and meat quality, and the birds are slaughtered in federally inspected plants. Game bird meat has a high-protein, low-fat content and a gamey flavor, and is considered a specialty item, but with a growing demand. Free-range and grain-fed production is preferred, as this retains the natural flavor, which is reduced or even lost when they are raised intensively on a diet of chicken pellets. A white ring-necked pheasant has been produced that gives a better appearance when dressed, as it is free from the black spots the dark feathers leave when the bird is plucked. The bobwhite quail is bred commercially in the United States for restocking shooting preserves.

The major grassland birds farmed commercially for their eggs, meat, oil, skins, and feathers are the emus and ostriches. Emus, farmed in many countries, are especially popular in their native Australia, including Tasmania where the island race was exterminated long ago. The goal of the industry in Western Australia is to produce 120,000 birds annually. Their popularity has boomed in the United States also, with about 1.5 million emus on 6,000 farms, supplying the growing market for meat and leather as well as oil for cosmetic use. In 1995 the U.S. Department of Agriculture approved meat inspection for all four grassland ratites—thereby also including the two species of rheas. Emu skins produce very high-quality leather, and the skins from captive birds are superior to those of wild emus, which suffer damage from the environment and during mating. But the ostrich is the favored ratite for farming.

Ostrich Chicks   Ostrich farming is now a major industry in many countries, the world's largest birds providing meat, skins, eggs, and feathers for commerce. Their chicks communicate while still in the egg and can synchronize their hatching, shortening their average incubation period of 42 days by several days, so they all hatch together.
Photo: Photos.com

Ostriches were used by the ancient Romans for riding and pulling chariots, but they were wild-caught birds from North Africa and Arabia, where they are now extinct. Regular captive breeding, and therefore domestication, began in Algeria in 1856, and in South Africa just a few years later, to meet the European demand for feathers for fashion, and before the end of the century farming was also established in North America, Egypt, Australia, and New Zealand. But these farms did not survive the First World War and the loss of the feather market, and the South African farms also dwindled then. The demand for ostrich meat and hides after World War II rekindled interest in farming in South Africa, and the increased demand in the 1980s encouraged farming elsewhere, including Britain, New Zealand, Australia, Zimbabwe, and the United States. Ostrich farms now produce many products, including meat, feathers, hides, eggs, and tendons that have application for human leg surgery. The use of ostriches' large eyes for human cornea transplants is being investigated, as are their brains for possible value in the treatment of Alzheimer's disease. Ostrich meat has a low-fat content and therefore less cholesterol, and is nutritionally similar to poultry and beef. Their hides provide supple and durable leather, less prone to drying and cracking and with a unique pattern caused by the widely spaced feather quills. They are in demand for high-fashion leather goods, comparable to those made from crocodile skins. The inevitable production of mutants has accompanied modern ostrich breeding, and albinos (called snow-flakes) and melanistic birds have been produced.

■ DOMESTICATING BIRDS

When captive birds breed regularly, they must eventually become domesticated. All aspects of their lives are totally controlled. Their mates are selected for them; their diet, housing, environment, and even the timing of their reproduction are all arranged; and they may be genetically manipulated. The great increase in bird breeding during the twentieth century, which has led to the production of new forms, resulted from major improvements to their diet and their environment, especially the increased knowledge of their breeding requirements and the ability to sex them. Of special mention are the softbilled birds such as toucans and turacos; many birds of prey, especially the falcons; and a host of parrots and their relatives. Other practices that improved breeding results, and therefore the beginning of domestication, included artificial selection, inbreeding, hybridization and the production of mutants, artificial egg incubation, and the fostering of their young.

Artificial Selection

Artificial selection, also called selective breeding, is best described genetically. The gene is the basic unit of inheritance. It contains the hereditary units that duplicate each time a cell divides, producing exact copies that determine the individual's characteristics. Genes may mutate, however, and produce a copy that is not exactly the same. Every bird gets two copies of each gene, one from its mother and one from its father. If a genetic trait is recessive, a bird needs to inherit two copies of the gene for the trait to be expressed. Thus, both parents have to be carriers of a recessive trait in order for the young to express that trait. If both parents are carriers, there is a 25 percent chance the offspring will show the recessive trait. The genetic mechanism that draws recessive genes out from the cover of the wild genotype of the natural species also brings about the first domestication-dependent changes. Nature actually has a store of various types and forms hidden as recessive mutants in every natural population of wild animals. It is this mutant pool that is exploited by humans, either purposely when the potential breeding pair are selected for a specific trait, or accidentally when a pair just look like nice birds. Such interference plays a creative role in developing new mutants of wild animals, or new breeds of domesticated ones.

Artificial selection is the opposite of the natural selection that occurs in the wild and results in the survival of the fittest and evolution. Whereas natural selection creates stabilized biological systems that ensure the development of the normal or wild phenotype, enabling it to adapt to changes and ensuring continuation of the species, artificial selection breaks down these systems, allowing gene combinations that may not survive in nature and providing a range of new possibilities. Hidden recessive genes in wild animals are exploited by captive breeding, especially by inbreeding, producing variations dependant on domestication, and providing the basis for change from the wild ancestor. Several factors affect wild nestlings' chances of survival, including lack of food, predation, and unfavorable weather, and these frequently cause the loss of some of the chicks in a nest. The survivors obviously had some advantage over those who succumbed, thus favoring their continuation and evolution. Removal of these risks happens all the time when birds breed in captivity, for every care is normally taken to ensure that the whole brood survives. There is really no alternative, for the breeder cannot assume the responsibility of natural selection. Consequently, birds with deleterious genes will be unavoidably raised. There may also be selection for certain characteristics that are advantageous for captive birds, such as increased tameness and ability to thrive on the replacement diet, that may eventually be controlled genetically. The very process of domestication obviously includes specimens that have a greater tendency to reproduce in man's care, thus involving selection even if it is unintended.

Inbreeding

Inbreeding is the mating of related animals that have one or more ancestors in common, the degree of inbreeding depending upon the closeness of the parent's relationship. Two very inbred parents that are unrelated to each other do not produce inbred young. Shared ancestry means that offspring could share genes that are the same copies of a single ancestral gene, which are said to be identical by descent. The main result of inbreeding is the reduction of individual animals that are heterozygous (possessing two different forms or alleles of a specific gene) and an increase in those that are homozygous (having two identical alleles of a particular gene), thus increasing the chances of mutants appearing. The degree of inbreeding is known as the inbreeding coefficient, and it can range from zero, when the parents are completely unrelated, to almost 100 through the continuous mating of closely related animals, especially brother and sister (known as full-sibling matings).

The loss of beneficial genes has a detrimental effect known as inbreeding depression, and this causes degeneration and infertility. It is unavoidable when only a small founder population of a species is available for breeding, and it actually occurred in several species of pheasants maintained by aviculturists. Edwards pheasant (which was believed extinct in its native Vietnam for many years until rediscovered recently), the eared pheasants, Hume's pheasant, and Sonnerat's junglefowl all became very inbred due to their small initial numbers and the impossibility of acquiring new blood. Wars, political boundaries, and then conservation and animal health legislation prevented the acquisition of new birds for many years. The captive population of the brown eared pheasant in the Western world stemmed from five birds imported into Europe in the 1860s, and new birds were not acquired from China until the 1980s.

A major difficulty in saving birds whose numbers have reached a critical low is their unknown genetic quality, especially the presence of deleterious recessive genes, which are then manifested by inbreeding. However, electrophoretic analyses (the separation of protein molecules with an electric current) seem to indicate that there is less genetic variation in bird populations than in mammals, and more inherent individual variation (birds are more heterozygous than mammals). The genetic character is better represented in a few specimens than in the same number of mammals, providing improved prospects of breeding from a small founder group. This was proven by the New Zealand scaup (*Aythya novaeseelandiae*) and the Laysan teal, both of which continued to breed well despite the small size of their founder populations. Even just two birds may be adequate to establish a thriving population; the 1,500 ultramarine lorikeets on the island of Ua Huka in the Marquesas stem from just one pair of birds liberated there in the 1940s. However, whereas inbreeding was unavoidable for some species involved in aviculture, it is now being practiced purposely by commercial breeders to create new birds.

Mutants

Mutants, morphs or aberrations, result from a permanent change in a cell's DNA sequence, caused by copying errors in the genetic material during cell division. This produces birds that differ from the wild or normal type of the species. Mutations affecting plumage color are a naturally occurring phenomenon in birds, resulting in the occasional appearance of such contrasting individuals as pied blackbirds and yellow (lutino) budgerigars. In the wild, these birds seldom survive beyond one generation because they are different. They lack the selection opportunities that are possible in captivity, and they are more noticeable and therefore more vulnerable to predators. Mutants also occur occasionally among confined birds, but the most successful method of producing them, rather than await their accidental arrival, is to allow closely related specimens to interbreed indiscriminately. When a mutant appears, its perpetuation, and of others stemming from it, are ensured through selective breeding. There is great interest in the production of color mutants, and some are indeed stunning birds. In its normal plumage, the ring-necked parakeet is a very attractive bird, pale green with a pink-and-black ring around its neck, but the same species in bright blue or yellow are equally lovely birds, favored by many people over the original natural color, and even considered an improvement over nature. The lutino is now a very common mutant, a lovely buttercup-yellow with pink eyes, a red bill, and a rose-colored collar. The blue mutant, in shades of powder blue with an off-white neck ring in the males (absent in the females), is also commonly bred. The albino ring-neck is snowy white with a pink beak and eyes and lacks a neck ring.

The production of mutants is widespread in the parrots and their relatives, and in some finches, and as breeding intensifies many others can be expected as a result of inbreeding and intentional selection. The breeding of color mutants and the development of new ones is now an integral aspect of the production of several species of parrots. It involves birds such as the quaker parakeet, now available in blue, lutino (yellow), albino (white), pied, and cinnamon mutants. There is a white-breasted mutant of the scarlet-breasted parakeet; and gray, albino, fallow (cinnamon with red eyes), khaki, and pied ring-necked parakeets are available. Bourke's parakeet has rosy, cinnamon, and yellow mutants, and the turquoisine parakeet can be acquired in pied, yellow, or red-fronted yellow forms. Lovebirds have also produced many color morphs. They have bred regularly since first imported from Africa in the middle of the nineteenth century, but there have been many imports since then, so it is unlikely that any pure lines survive from the original imports. However, even with the injection of new blood over the years, many color mutants are now available, and they are considered thoroughly domesticated. The peach-faced lovebird is available in lutino, pied, blue, dark green, and white mutants. There are blue, yellow, and white masked lovebirds, and Fischers's lovebird has produced blue and yellow variants.

These color mutants occur as a result either of changes in the structure of the genes, or of genetic defects. Blue-colored birds result when a genetic mutation prevents them from synthesizing red and yellow pigments, thus converting their normal green color to blue. The bright yellow lutino is a bird that cannot synthesize melanin but still produces yellow and orange pigment. Domestically perpetuated mutants of cockatiels all result from changes in the formation of melanin, with a different result in each genetic mutation—lutinos, pieds, cinnamons, pearls, and fallows.

Ornamental pheasants have always been attractive to aviculturists. They are mostly colorful species and include several of the world's most beautiful birds, especially the golden pheasant, the Lady Amherst pheasant, and the tragopans. With few exceptions they are easy to maintain and breed; most species have reproduced continually for over a century and must be considered domesticated. Most are in trouble in the wild, but since they were first exported from Asia in the middle of the nineteenth century, the trade in wild-caught birds has been minimal and certainly cannot be blamed for assisting their demise. In fact, recent surveys have shown that there are more pheasants currently in aviaries than have ever been exported from their countries of origin in a century and a half. The greatest chance of survival for several species now depends upon captive-breeding, but the problem is the difficulty of acquiring new blood for the long-term strains descended from the few original or founder birds. Despite their long history of domestication, color mutants are not common in the galliformes. Melanistic and golden-buff mutants of the long-domesticated ring-necked pheasant have been bred, and several color and pattern mutants of the peafowl have been produced also. But none of these are an improvement over their wild ancestors. Of the more recent domesticates, the golden pheasant is available in salmon, yellow, and dark-throated versions. The lovely yellow-golden mutant was bred by Dr. Alessandro Ghigi of Bologna in 1952.

Several small seed-eating birds have been bred so regularly they must also be considered domesticated. The greenfinch (*Carduelis chloris*) has been bred in sufficiently large numbers to select and establish color mutants, including cinnamons and lutinos. The beautiful Gouldian finch (*Chloebia gouldiae*) is available in an incredible array of colors, and so many mutants of the zebra finch have been developed that the Zebra Finch Society has exhibition standards for judging them. Several species of small doves have also been bred for many generations. The barred dove (*Geopelia maugie*) and zebra dove (*G. s. striata*) are very common in aviculture, and the free-breeding habits of the pretty little diamond dove (*G. cuneata*)—the smallest pigeon—have resulted in the production of numerous color mutants, including blue, silver, cinnamon, and yellow. The larger members of the genus *Streptopelia*, all characterized by their dark collar or scale-like patches of darker feathers on their necks, have bred readily in aviaries. Several species of turtle doves, and seven species and many races of birds known as collared or ring doves, also breed prolifically; and one has produced a domesticated form known as the Barbary dove, whose ancestors were either the Asian collared dove (*S. decaocto*) or the African collared dove (*S. roseogrisea*), which has self-populated Europe. The Barbary dove is available in several mutants, of which the albino has been popular for many years and is often called the Java dove. A silky form is also an hereditary strain.

Hybridization

Hybridization is the cross-breeding of birds of different species (interspecific hybridization) and of birds of the same species but of different subspecies or races (intraspecific crosses). Whereas offspring stemming from intraspecific hybridization are fertile, fertility of the young from interspecific crosses depends upon the degree of relationship between the species. They may be fertile but with reduced fertility (males fertile, females infertile) as the genetic differences become wider apart, until a point is reached where the offspring of both sexes are sterile. In matings between distantly related species, such as the chicken and the turkey, or the budgerigar and lovebird, they may not produce a fertilized ova, or the embryo dies before hatching, which is nature's way of keeping the species pure. Therefore, hybridization confirms a genetic relationship between species if offspring are produced, and a closer one if they are fertile. Examples of viable interspecific crosses are the golden pheasant with the Amherst pheasant, the splendid grass parakeet with the turquoisine grass parakeet, and crosses between waxbills of the genus *Estrilda*. In such cases, hybrid vigor may result in young that are more viable than those from a pure specific mating. The offspring of more distantly related species such as a canary crossed with a red siskin, which are hybridized to produce red factor canaries, are less fertile. Hybridization is not restricted to captive birds and occurs naturally in the wild, especially in subspecies whose ranges overlap, but also irregularly between species, and may have a role in the evolution of new forms.

Hybridization in aviculture was originally mostly accidental, resulting from the housing together of subspecies or related species due to identification errors or the lack of mates or space. Now it is practiced to produce birds that may have commercial value, out of an interest in genetics, or simply out of a desire to be the first to produce a particular hybrid. It is both a natural cause and effect of domestication and therefore must increase as bird breeding intensifies. Deliberate hybridization has occurred in many bird families, including pheasants, waterfowl, and finches, followed by selective breeding to enhance or delete certain characteristics, and some of the resulting offspring are very attractive birds. It has been particularly common in the parrots and their relatives, and many interspecific hybrids have been produced. Lovebirds hybridize freely; the masked lovebird has mated with both the peach-faced and black-winged lovebirds. Black-cheeked lovebirds have hybridized with Nyasa lovebirds, and Fisher's lovebirds with peach-faced lovebirds. The first blue Fischer's lovebirds resulted from crossing the normal type—a green bird with orange-red face and throat—with the blue mutant of the masked lovebird. Among the Australian parakeets, there are also many examples of interspecific crossing—for example, the red-rumped parakeet's willingness to breed with members of the same genus (*Psephotus*), such as the many-colored parakeet and blue-bonnet parakeet, and with several species of rosella parakeets (*Platycercus*). All the species of grass parakeets (*Neophema*) hybridize interspecifically with the exception of Bourke's parakeet. The three members of the genus *Polytelis*—the barraband parakeet, rock pebbler, and the Princess of Wales parakeets—have all bred interspecifically, and the barraband parakeet has also crossed with birds of different genera, such as rosellas (*Platycercus*) and king parakeets (*Alisterus*).

Closely related species of pheasants also readily hybridize, and have been allowed to do so, to the extent that many impure specimens exist. Intraspecific crosses are common, such as the subspecies of the kalij pheasant that have been so frequently hybridized they are difficult to identify. Examples of common interspecific crosses producing fertile chicks are the golden pheasant and the Lady Amherst pheasant; the silver pheasant and the kalij pheasant; and the brown eared pheasant with the blue eared pheasant. They may also produce fertile young when crossed with species from other groups. For example, the golden pheasant has crossed with Reeve's pheasant and with the silver pheasant, in which the males are fertile but the hens are sterile. If the objective of hybridizing is to eventually produce new "breeds," then of course crosses resulting in sterile young of both sexes, such as the mating of a golden pheasant and a cheer pheasant, are of no value. Other gallinaceous or chicken-like birds that have hybridized include the ocellated turkey of Central America with the wild turkey of North America. Most of the red junglefowl in aviculturists' aviaries may be impure, and many of the wild populations have been contaminated by chickens. In some parts of the junglefowl's natural range, both chickens and hybrid junglefowl have reverted to a semiferal existence and breed with the wild red birds. Pure wild males have an eclipse plumage (dull feathering acquired at the end of the breeding season) that has not been observed in the wild populations for many years (for over a century in the Philippines), indicating they have probably been genetically contaminated. Maintaining pure species of all birds in captivity, in addition to the production of new birds, is therefore very important, and serious aviculturists must make every effort to maintain the purity of their stock.

Hybridization has also been employed in an attempt to recreate a species. The imperial pheasant is a very rare species found only in Laos and Vietnam, of which all the captive birds stem from a pair collected by Jean Delacour in 1923. Just one bird survived World War II at the Antwerp Zoo; it was crossed with a silver pheasant hen, with selective hybridizing then practiced for many years to produce birds resembling the original wild species. Wild pheasants were captured in Vietnam in 1990 and 2000, and from these it has since been determined that the imperial pheasant is actually not a true species after all, but a naturally occurring hybrid between Edward's pheasant and the silver pheasant.

In addition to the production of pure birds of prey, the fact that several diurnal raptors are closely related has also resulted in considerable hybridizing between species and subspecies within the genus *Falcon,* and to a lesser extent in the hawks. Some have bred naturally, but artificial insemination has also been widely used. The falcons are so closely related that many fertile crosses have been achieved, and hybridization continues in an attempt to produce birds with characteristics of more value to the falconer, combining the best features of their parents. Captive-raised birds are now generally believed to surpass wild-raised birds for falconry, and are preferred in Arabia. The peregrine falcon has been crossed with the merlin, gyr falcon, Barbary falcon, lanner falcon, saker falcon, and prairie falcon. The hybrid peregrine-merlin is a very fast flyer and willing to circle and wait for the quarry to be flushed, which pure falcons are usually not prepared to do. The gyr falcon has also hybridized with the merlin, saker falcon, and prairie falcon; and the kestrel has crossed with the prairie falcon. Many Harris hawks, considered the best hawk in contemporary falconry, have been bred, and hybrids between it and the red-tailed hawk have also been produced. The production of domestic "breeds" of falcons is therefore occurring at a much faster rate than the original domestication of the chickens, ducks, and geese. Hybridization allowed breeders to circumvent the restrictions placed upon falcon possession and sale, for the hybrids can be legally kept and sold as they are neither one species nor the other.

Ducks hybridize more readily than other birds, and this occurs frequently in free-living populations, a fact that is obvious on lakes in public parks literally anywhere in the world. Several hundred hybrids have been recorded, most in captive collections but many in the wild, and the ability and readiness of waterfowl to accept mates of other species is considered a serious problem. The mallard readily crosses with many other ducks, including the gadwall, pintail, shoveller, and wigeon, and the hybrids have then mated with other species including the tufted duck, canvasback, and redhead, making identification very difficult. The wood duck is also a frequent hybridizer. A species that is threatened by hybridization in the wild is the globally endangered white-headed duck (*Oxyura leucocephala*), which, possibly due to its small numbers, often mates with the ruddy duck (*Oxyura jamaicensis*). In New Zealand, the introduced mallard has hybridized with the native gray duck (*Anas s. superciliosa*), and pure gray duck genes now exist in only 20 percent of the population. A captive life involving segregation may be the only way to ensure the continued purity of some species.

Glofish   Developed for the ornamental freshwater fish trade from the familiar and popular zebra fish, the beautiful fluorescent glofish was the first genetically modified animal available as a pet. It is now produced in three colors, red, green, and orange, which glow at night when their tank is lit with a black light.

Photo: Courtesy www.glofish.com

Pigeon Blood Discus   This very pretty fish, a mutation of the blue discus, was developed from a single mutated fish in which the normal vertical black bars were absent, making the fish appear much brighter. The females cannot darken when spawning, which they normally do to attract the fry to feed on their skin excretions.

Photo: Courtesy Anka Zolnierzak, Wikipedia.com

Super Hypo Tangerine Leopard Gecko   Leopard geckos have been popular pets, and important laboratory animals, for many years, and have proved easy to breed in small plastic containers. Normally cream-colored with black spots, selective breeding over many generations has produced dozens of color mutants, none more spectacular than this one.

Photo: Courtesy Steve Sykes, steve@geckosetc.com

Dyeing Poison-dart Frogs   A stunning species from the Guianas, it was apparently named as a result of being used to dye fabrics. Now being captive-bred, breeders provide their frogs with a "hut" made from a halved coconut shell, with a tiny dish of water inside in which to lay their eggs.

Photo: Photos.com © JupiterImages Corporation

Emerald Tree Boa An arboreal live-bearing snake, like the other boas and unlike the pythons, the emerald tree boa has proved more challenging to breed than many of the more commonly kept boas, but is now beginning to breed regularly. The many pattern and color variations that occur naturally over its wide range in Amazonia and the Guyana Shield will aid the development of captive strains.

Photo: Stephen Bonk, Shutterstock.com

Albino Burmese Python This startling mutant of the normal dark-brown Burmese python was the first to be regularly produced by herpetologists. Many other morphs are now available, including snakes in which the regular pattern has been replaced with speckles.

Photo: Ishbukar Yalilfatar, Shutterstock.com

Feral Guineafowl *Originally kept in Ancient Rome and Greece, the guineafowl was apparently lost as a domesticated bird until redomesticated in the fifteenth century when Portuguese traders brought them back from West Africa. It once again became a favorite for the table and as a farmyard bird. Escapees now live ferally in Jamaica, Australia, and New Zealand.*
Photo: Clive Roots

Hyacinthine Macaws *The largest of the macaws, the hyacinthine was the most recent species to begin breeding regularly, due more to its rarity in collections than for any other reason. Like several of the large macaws, it is now produced regularly by pet trade breeders. Poor talkers, and rather noisy, hand-raised chicks are very gentle birds, but their large and powerful bills demand a very sturdy metal cage.*
Photo: Bruce Wheadon, Shutterstock.com

Wild Ring-necked Parakeets, Cleartail Turquoise-gray Ring-necked Parakeet, Pallid Turquoise Ring-necked Parakeet, Violet Ring-necked Parakeet

Many attractive mutants of the "normal" wild ring-necked parakeets (top left) have been produced by breeders. Three of the rarer forms are (clockwise from the top right) the clear-tail turquoise-gray, the pallid turquoise, and the violet.

Photos: Courtesy Z. Rana. www.psittaculaworld.com

Eclectus Parrot   One of the most spectacular parrots, now being produced regularly by aviculturists and breeders supplying the pet trade. Contrary to the general rule in birds, the hen (above) is more colorful than the bright green male. Eclectus parrots are not good talkers, but hand-raised babies are affectionate and intelligent birds.

Photo: Nicola Gavin, Shutterstock.com

Sulfur-breasted Toucan   The breeding of toucans was one of the major developments in aviculture in the latter half of the last century. The commonly kept Central American sulfur-breasted toucan, also known as the keel-billed toucan, was first bred in 1976, and has since reproduced regularly in zoos, bird gardens, and the collections of private aviculturists.

Photo: Michael Strzelecki, Shutterstock.com

Wild Boar Piglets   Pure wild boar are now farmed in several countries as a source of lean meat. Escapees are already established in the wild, in some places joining their long-domesticated descendents, the white farm pigs, that are already feral. The piglets of pure wild boars have striped coats like the ones above.

Photo: Roland Syba, Shutterstock.com

North Chinese Leopards   Now exceedingly rare and fragmented in their native China, actually in the central districts despite their name, and with only about 60 surviving in zoos outside China, the low founder population of this leopard subspecies makes its continued reproduction, and therefore its domestication, an unlikely prospect.
Photo: Clive Roots

Feral Dromedaries   Camels were used in the Australian desert in the nineteenth century as beasts of burden, but with the coming of roads and the railway, they became redundant and were abandoned to their fate. They thrived in the outback, having evolved in a similar habitat, and now number about half a million.
Photo: Styve Reineck, Shutterstock.com

Bengal Cat   This very attractive cat is the product of selective breeding involving several breeds of domestic cats and the wild Bengal leopard cat. As it contains the genes of both domestic and wild animals it cannot really be called a breed or a species, so it is called a "designer breed." It has been bred in a wide range of colors and patterns.

Photo: Marilyn Barbone, Shutterstock.com

White Tiger   Captive white tigers, which are not albinos, are descended from a rare mutant Bengal tiger caught in India in 1951. They have been perpetuated, even in some of the world's major zoological gardens, through inbreeding and possibly hybridization, and are therefore considered "generic" tigers.

Photo: Photos.com © JupiterImages Corporation

Fostering

The use of foster parents has also aided the captive breeding of birds and therefore their domestication. Fosters are used when birds have difficulty incubating their own eggs or raising their offspring, and when indeterminate layers, which are those that continue to lay when their eggs are removed, lay many more than their normal clutch and cannot incubate them. Bantams are the most frequently used fosters for birds the size of pheasants and large quail, whereas the eggs of finches that cannot be trusted to incubate their own eggs or raise the nestlings are given to the long-domesticated Bengalese finch (*Lonchura domestica*) to incubate. Zebra finches (*Taeniopygia guttata*) are often so keen to continue laying eggs that they put more grass on top of their eggs and lay another batch, so their eggs may also be given to the very willing Bengalese. Captive wild pigeons, such as the wonga-wonga pigeon (*Leucosarcia melanoleuca*) and the crested pigeon (*Ocyphaps lophotes*) are sometimes uninterested in proceeding beyond the egg-laying stage, and domestic pigeons and barbary doves are used as fosters. Most parakeets are very receptive to hatching the eggs of other species, and even to fostering chicks that have been hatched artificially—in an incubator. Red-rumped parakeets (*Psephotus hematonotus*) are very reliable fosters and have raised the chicks of many other species, including rosellas and crimson-winged parakeets (*Cryptospiza reichenovii*). Jenday conures (*Aratinga jandaya*) have fostered other conures, and the smaller species of Amazon parrots. Some breeders believe it is better for the chicks to have a natural rather than a fostered upbringing, as the parents may lose their chick-raising behavior. However, some have lost this already or they would raise their chicks, and for others it would be just one more domestication-produced change. It could have serious consequences for reintroduction, but foster-raised birds and those with a history of chick neglect are unlikely to be considered for release anyway.

Artificial Incubation

Artificial incubation is now a common avicultural technique, in which incubators rather than foster birds are used to hatch eggs. For birds that cannot be relied upon to incubate their own eggs or raise their young, for the ratites that lay many more eggs than the male bird can cover, and when the climate or companions jeopardize the safety of the eggs or resultant young, eggs are artificially incubated. Also, commerce and to a lesser degree conservation now dictate the removal of eggs to stimulate increased egg production, and to avoid wasting a hen's time and energy in such lengthy natural tasks as incubation and chick raising. Artificial incubation is an aspect of modern bird-keeping that greatly contributes to domestication, for hatching eggs artificially involves either the hand-raising of altricial chicks or the very considerate care of precocial ones. Consequently, they are not subjected to the pressures, including competition from siblings and possible exposure to the elements, to which naturally raised chicks are subjected, when the weaker individuals may be eliminated. Also, parent-hatched and raised young are assumed to have advantages over captive-raised chicks, at least for reintroduction, as imprinting to the parent results in more natural behavior that is of greater value in the wild. The chicks of some species imprint readily on their keeper, which

Black-headed Caique   *A medium-sized parrot from northern South America, the black-headed caique is now bred regularly by pet-trade suppliers. Hand-raised individuals are charming birds, playful, inquisitive, and very active, but they are not good talkers.*
Photo: Kevin Britland, Shutterstock.com

must be avoided through the use of hand puppets created in the likeness of their natural parents.

Problems

The domestication of new birds in the twentieth century also had some unexpected and unwanted results. Inbreeding depression was one of the most serious, and of course to be expected, due to the repeated breeding of closely related individuals. It is caused by the deleterious effect of recessive genes, resulting from the increase of homozygotes until they eventually outnumber the heterozygotes as inbreeding continues. Inbreeding depression is manifested by unhealthy animals, with poor reproduction (due to low fertility and poor viability of the young) and increased susceptibility to infections. Budgerigar breeders occasionally experience a serious condition in their flocks, in which the chicks are unable to grow their feathers, and consequently are called "runners." It is known as French moult, so named because it was first seen in Europe in the middle of the nineteenth century, and has been blamed on mites that destroyed the feathers, on a bacterium and a mycotoxin, also on diet and overbreeding, but the cause is still unknown. A similar feather problem has occurred in cockatiels, probably as a result of inbreeding; that has resulted in baldness, which first appeared in the 1950s and is still experienced by breeders today. Hereditary behavioral defects in the early days of lutino cockatiel production resulted in infertility when males attempted to mate while facing the wrong direction. Domesticated pigeons have also suffered from a condition known as hereditary ataxia, a genetic disorder resulting from inbreeding and the effect of recessive genes. Affected birds are uncoordinated, stumble about, fall over backwards, and fly poorly.

■ SOME OF THE SPECIES

Hawaiian Goose (*Branta sandvicensis*)

This is the world's rarest goose, which was almost exterminated in the 1940s. A native of several of the Hawaiian Islands, it was named the state bird of Hawaii in 1957. Also known as the nene, the Hawaiian goose is adapted for life on the rocky, sparsely vegetated volcanic lava fields and rarely enters water. Consequently, it has almost lost the webbing between its toes. Once a common bird, hunting and introduced predators such as mongooses, feral cats, dogs, and pigs reduced its original population of 25,000 to just 25 birds in 1950. Its decline actually began with the arrival of Europeans on the islands in 1778. The geese were hunted, their eggs were collected, they were killed and salted for provisioning whaling ships, and they were even shipped back to California to feed the gold miners. The arrival of the predators, especially the mongoose in 1883, sealed their fate. The Hawaiian goose is smaller than the Canada goose, about 24 inches (60 cm) long, and is unmistakable with its black head and large buffy-yellow cheek patches. All the existing stock are derived from the few birds caught and bred in captivity, firstly by Herbert Shipman on his property at Haena Beach on Hawaii, and by the Wildfowl Trust in England. To save the remaining birds, The Hawaiian Goose Recovery Plan, published in 1983, called for the establishment of 2,000 nene on Hawaii and 250 on Maui. Many captive-raised birds have been reintroduced, but they are still threatened by mongooses, rats, and feral pigs, cats, and dogs, plus low productivity resulting from the small founder stock and the initial inbreeding to save them. Maintaining the wild population is also still dependent upon continued releases of captive-raised birds, and since 1960 2,500 geese have been released on Hawaii, Maui, and Kauai, but the droughts between 1976 and 1983 killed most of the birds released in those years. In 2004 there were estimated to be 500 birds in the wild, on Hawaii, Maui, and Kauai, and about the same number in the collections of private breeders and zoos.

Blue and Gold Macaw (*Ara ararauna*)

A large and very colorful member of the parrot family, the blue and gold macaw is one of the most popular and readily available species. For the past three decades it has been bred consistently, and ads in pet magazines and on the Internet list dozens of captive-bred and hand-raised babies for sale. A native of tropical America, from Mexico to northern South America, the blue and gold macaw is about 35 inches (90 cm) long and has a long, typical macaw tail. Its back and upper tail feathers are brilliant blue, and the under tail feathers and breast are orange-yellow. It has a green forehead and its wings are blue with green tips. Typical of the macaws, it has a mostly bare white face with lines of tiny feathers beneath the eyes, a large and powerful beak, and a thick, fleshy tongue. Like all parrots, the macaws have zygodactylous feet, in which two toes face forwards and two to the rear, which allows great dexterity for climbing and holding their food. Like all the large macaws, they are noisy, destructive birds and require metal and mesh cages. Exposed woodwork is soon destroyed, and even the entrance holes to their nest boxes must be protected with sheet metal. In the wild they nest in tree cavities, laying two eggs on the rotting

Hawaiian Goose or NeNe   *The Hawaiian goose evolved on the volcanic lava fields, but the European colonization of the islands resulted in its virtual demise, due to overhunting and predation by the numerous introduced predators, by 1950 only 25 geese survived. Captive breeding has since produced 2,500 geese for reintroduction, but their survival rate has been low.*
Photo: Courtesy USFWS, photographer Glen Smart

wood at the bottom of the hole. They are monogamous, and the male feeds his mate while she sits on the eggs for 25 days. The newly hatched chicks are altricial—naked, blind, and helpless—and are fed by both parents with regurgitated fruit and seeds. Captive-hatched chicks are normally removed for hand-raising when they are several days old, as this allows the parent birds to pass on the intestinal flora important for their digestion, and may reinforce the chick's own parental behavior when it eventually breeds and raises its own young. A typical traditional macaw diet includes fresh fruit and vegetables, peanuts, and sunflower seeds, plus monkey chow or dog chow; but commercially produced "complete" pelleted diets are now available, scientifically formulated to provide the bird's total nutritional requirements.

Ring-necked Parakeet (*Psittacula krameri*)

The ring-necked parakeet's range is a wide band of country north of the equatorial forests in Central Africa from the Atlantic Ocean to the Red Sea, and southern Asia from Pakistan to Myanmar south of the Himalayas. It prefers dry forest and wooded grasslands but has successfully combated habitat destruction and urbanization and is at home in "disturbed" habitat, and in plantations, parks, and gardens. There are four subspecies, two in Africa and two in southern Asia, and the latter birds have been introduced and are established worldwide, mainly as a result of escapee pets. Also known as ringnecks, they are now established as feral birds in Florida, California, and southern England, and live in several European cities. The ring-necked parakeet's natural diet is seeds and grains, fruit, nuts, buds, and shoots. It is a noisy, communal species, living in flocks until pairs separate to nest, when they find a natural tree cavity, an old woodpecker hole, or a hole in a wall.

Both parents incubate the two to four eggs for about 23 days, and the chicks take six weeks to fledge.

This parakeet is one of the most prolific breeders of all the small parrots, and during the last century has become a favorite aviary bird also because of its hardiness and adaptability. In its natural coloration it is a very attractive bird; pale green with black wing feathers, a long, tapering bluish-green tail, a large red beak, and a pink and black ring around its neck. It is about 16 inches (40 cm) long, including its tail. The production of many very attractive color mutants has increased its popularity. They include the commonly produced lutinos and blues, plus albinos, turquoise, cinnamon, pallid-ino (originally called yellow-headed cinnamon), pallid (formerly called lacewing), and the turquoise lacewing. It has also hybridized frequently with the alexandrine parakeet (*Psittacula eupatria*), and several mutants have resulted from this hybridization; but fortunately they are not always fertile. It has also crossed with the slaty-headed parakeet (*Psittacula himalayana*) and the moustached parakeet (*Psittacula alexandri*).

Peach-faced Lovebird (*Agapornis roseicollis*)

Lovebirds are small, mostly green parrots with a colorful head or breast. They hail from Africa, Madagascar, and several offshore islands, and their name stems from their mutual preening behavior and their loving attention to their mates. Several species have been popular cage birds for many years, and are now thoroughly domesticated and available in numerous very attractive color mutants. The development of new forms has been aided by the fact that the species readily hybridize, and to maintain their purity they must be kept apart. The peach-faced lovebird is a native of southwest Africa, where it lives in dry scrub and riverine forests. It is the largest lovebird and the most popular, an affectionate, playful, and intelligent bird. It is 6 inches (5 cm) long, and its green body plumage is enlivened with a rose-pink forehead, face, throat and upper breast; and it has a bright blue rump. Although they are colonial, beyond their loving pair relationship lovebirds are bad-tempered and quarrelsome, and colonial life in captivity is impossible unless they have large aviaries and a surfeit of nest boxes. Pairs prefer to nest alone, and if the same species are housed alongside each other in small cages and have visual contact, the cocks will fight through the wire and jeopardize all chances of breeding.

Like most parrots, lovebirds nest in cavities, but unlike the others they actually build a nest of tightly packed twigs within the cavity. They are one of the few parrots to carry material to their nest hole, and most species carry twigs in their beaks or tucked into their body feathers, but the peach-faced lovebird actually tucks them into its rump feathers. Unfortunately, peach-faced lovebirds are monomorphic (the sexes are alike) so they must be sexed by laparoscopy, or by the latest method—DNA-based sexing using blood or feathers—that has achieved results with 99.9 percent accuracy. They have been bred in an amazing array of colors, with perhaps 16 primary color mutants and many color combinations or secondary mutants. They include the pied, the white faced, Dutch blue, cobalt, green violet, American cinnamon violet, and the lutino, a lovely golden-yellow bird with rose-pink face and upper breast. They have hybridized with the masked lovebird, Fischer's lovebird, Nyasa lovebird, and the black-cheeked lovebird, in each case producing some fertile offspring that could then be mated back to one of the parents.

Cockatiel (*Nymphicus hollandicus*)

The cockatiel is a common bird in Australia, where it lives in the temporary grasslands that appear only after rain, open savannahs, and the mallee or scrub desert. But its distribution is controlled by rainfall, and its breeding cycle is triggered by the onset of the rains. Its streamlined shape and well-developed wings allow fast migration over long distances in search of water and grass seeds. In the wild, cockatiels remain paired throughout the year and mating occurs regularly. Because of the state of readiness of their ovaries throughout the year, ova being developed and reabsorbed regularly, the cockatiel can lay fertile eggs just five days after a nest site is selected following a rainstorm. The wild cockatiel is 12 inches (30 cm) long including its tail, and has a black bill, feet, and eyes. Basically dark gray in color, the adult female has light areas on the underside of its tail and yellow bars on the wings, a dark gray head, and a pale orange cheek spot. The adult male is darkly colored under the tail and lacks wing bars, and has a yellow face, cheeks, and crest, paling behind the eye, with an orange cheek spot and a fuller crest than the female.

Although the cockatiel was first exported from Australia in the middle of the nineteenth century, the limited breeding and continued availability of wild birds until well into the last century did not result in mutants as early in its captive history as the more popular budgerigar. The first mutant, the pied, did not appear until over a century after the cockatiel's first importation into Europe. The modern cockatiel, and all its mutants, is therefore very much a product of the twentieth century. Like the budgerigar, it was for many years considered a beginners bird, a bird for the pet keeper—either as a single bird indoors in a cage, or a small flock in an outdoor aviary. Increasing interest in the cockatiel after World War II, coupled with the Australian bird export ban of 1959, resulted in selective breeding, and birds differing in plumage from the normal wild cockatiel began to appear. The harlequin or pied was the first of these mutants to appear just after the war, followed by cinnamons in the early 1950s and lutinos in 1958. The availability of these attractive mutants increased interest in cockatiel breeding, especially in view of the initial commercial value of the mutants, and during the next decade many breeders concentrated on finding and perpetuating mutants. In Europe new forms such as the pearl or laced cockatiel were bred in 1967 and the red-eyed or fallow in 1968. So many birds were produced that the market was soon flooded, and the formerly valuable mutants were soon available for little more than the cost of the normal wild birds.

The cockatiel mutants all result from changes in the formation of melanin. In the cinnamon cockatiel, the melanin granules are brown instead of black and do not absorb the same amount of light, resulting in a pale-brown appearance. The fallow or red-eyed cockatiel is a pale silver color as a result of reduced melanin deposition. In the pearl or laced cockatiel, the feathers of the nape, mantle, and upper wing coverts are the only ones affected by the bird's inability to produce melanin in the center of each feather, resulting in a pale patch on each side of the shaft and an overall scalloped or marbled appearance. The yellow base color is golden yellow in some birds, and in contrast with the gray is very attractive. They are also known as gold-laced cockatiels as opposed to silver-laced, in which the background yellow coloration is quite pale. Although normally colored cockatiel's eyes have a thick layer of melanin that absorbs the light entering the eye, both the lutino and albino cockatiels lack melanin and have sensitive eyes and poor sight.

Budgerigar (*Melopsittacus undulatus*)

Unlike all other pre-twentieth century domesticates, the budgerigar's history as a pet bird is recorded from the arrival of the first live birds from Australia. This happened in 1840, when the famous bird artist John Gould returned to England after traveling across Australia, researching for his book on the birds of the continent. The London animal dealer Jamrach sold the first pair of budgies for £271 ($570), and they became so popular that thousands were exported over the next few years, but they bred so prolifically that within two decades they were cheaper to buy in England than in Australia. They proved to be the most popular cage bird ever, known as budgies in England, but often called parakeets in the United States

The wild budgerigar (a corruption of their aboriginal name, betcherrygah) is a common bird in interior Australia. It is gregarious and nomadic, living in large flocks that move continually in search of water and seeding grasses. The flocks disappear for months from a region and return suddenly when heavy rains have turned the scrub and desert into a sea of kangaroo grass that seeds just six weeks after germinating.

Budgies have benefited from the water troughs provided for domestic stock, and although they can survive for up to four weeks on a diet of dry seeds alone, they suffer severely during long droughts. They are adaptable and undemanding, able to mimic a few words, and with human company they can live an otherwise solitary life in a cage for many years. The wild budgerigar is light green or dark green; yellow birds are occasionally seen, but as with all large animal populations the mutant genes are soon absorbed by the masses. However, the mutants that appeared among captive birds were perpetuated, and others were created by inbreeding, and budgies are now available in a bewildering range of colors. Yellow (lutino) birds first appeared in Belgium in 1872; sky blues early in the last century, also in Belgium; and then the first white budgies with dark eyes were bred in 1920 in England and France. The color standards of Britain's Budgerigar Society currently include opaline, white, sky blue, gray-dominant pied, and spangle light green birds. The budgies being bred for exhibition are larger than their wild ancestors, the Society's show standards calling for the ideal budgie length to be 8½ inches (22 cm), 1½ inches (4 cm) longer than the wild bird.

Edwards Pheasant (*Lophura edwardsi*)

This small pheasant is a member of the gallopheasant group of 10 species within the genus *Lophura,* which includes the very common and popular silver pheasant. It was first discovered in 1895, but all the captive birds in the aviaries of aviculturists, zoos, and bird gardens are descended from 15 pheasants collected and shipped to France by Jean Delacour in 1923. They were first bred in 1925, have not had any infusion of new blood since then, and have produced many generations, so they are indeed new domesticates of the twentieth century. The species was not seen again in the wild—the rain forest of Vietnam—for many years and was considered extinct, but was seen again recently in central Vietnam. It is believed its rarity stems from habitat disturbance, hunting, and of course the effects of war. Although it has reproduced well in aviaires, and has always been fairly common, its genetic diversity is very low, its captive population having started from so few birds years ago. An international studbook has been initiated for the species, and genetic research is currently being undertaken to determine the size of the gene pool, and for use as a tool for maintaining genetic variation. The male Edwards pheasant has body plumage of several shades of metallic blue, and a short white crest, bright red facial wattles, red feet, and a short tail. Hens are dull brown, with no crest and small pale wattles. Young males resemble hens until they get their adult plumage towards the end of their first year. This is one of the pheasant species that is fertile before it is one year old, and lays four to seven eggs that hatch after 23 days incubation. Edwards pheasants are omnivores; their natural diet is seeds, grain, berries, and invertebrates, and with their typical strong chicken-like legs they scratch in the leaf litter for much of their food. Captive birds, like most pheasants, receive a basic diet of chicken pellets, supplemented with grain and green food.

Red Siskin (*Carduellis cucullata*)

The red siskin is a small gregarious passerine or perching bird, about 4 inches (10 cm) long. For many years it was believed to be confined to northern Venezuela and Colombia, but recently a population of several thousand was discovered in southern Guyana, hundreds of miles away. It is a seed-eating bird of the open scrub–grassland and forest edges. The male siskin is deep red with a black head, throat, and wing feathers, and has a whitish belly. The female is gray and black. The siskin was common until the beginning of the last century, when it was trapped in large numbers for the pet trade. The demand stemmed from its ease of breeding and its willingness to hybridize with the domesticated yellow canary, to produce offspring with deep red plumage, known as red factor canaries. German breeders at the time believed that color was determined solely by inherited factors, so they were surprised to breed orange-colored canaries. It is now known that such inherited traits in orange and yellow birds also need carotene to produce deep red feathers, and breeders now give their birds canthaxanthin, a synthetic carotenoid pigment, that is also given to roseate flamingoes and scarlet ibis to maintain their color.

The demand for siskins led to their virtual extinction, but they are now strictly protected, although still very rare and occurring only in fragmented groups. They are one of the few birds directly endangered by the cage bird trade, and the loss of their genetic base through hybridization. Most captive red siskins are kept in Europe, particularly in Germany, but pure-blooded birds are now rare because breeders hybridized their birds with the green siskin to produce hardier individuals that could be housed outdoors. The American Federation of Aviculture has initiated a program to breed siskins for their eventual release in Venezuela; a studbook is maintained, and inbreeding and pairings are recorded.

Sulfur-breasted Toucan (*Rhamphastos sulfuratus*)

A major development in aviculture in the latter half of the last century was the breeding of toucans and their smaller relatives the toucanets and aracaris. The first breeding of the sulfur-breasted or keel-billed toucan occurred at the Houston Zoo in 1976, but it has been bred regularly since then by other zoos and several private breeders, notably California's Emerald Forest Bird Gardens, a leader for many years in breeding toucans and their relatives. It is a very colorful species, a black bird with a sulfur neck and chest; the base of its tail is red, and it has blue feet. But the large and striking bill is its most dominant feature; pale green with a bright red tip and orange blazes on the sides. It is very light despite its size, about 6 inches (15 cm) long. The sulfur-breasted toucan is primarily a fruit and berry eater, but may also eat the nestlings of other birds. It picks up food with the end of its beak, tosses it into the air, and catches it in the back of its throat to swallow it. Captive diets for the toucans usually include meat and dog chow in addition to fruit, but foods containing iron have produced fatal iron-storage disease (hemochromatosis). As iron is absent from the toucan's natural diet, they did not evolve a means of eliminating it when ingested, and it accumulates in their livers. All toucans nest in cavities high in rain forest trees, and captive birds have always been provided with nest boxes, but the major breakthrough in their breeding began when they were given old palm trunks, standing upright, with nesting cavities chipped out for them. The sulfur-breasted toucan lays one to four eggs and incubates them for 18 days. The hatchlings are naked and helpless and stay in the cavity until they fledge, at the age of about six weeks. It is a common species in the rainforests of Central America and northern South America.

Peregrine Falcon (*Falco peregrinus*)

A superb raptor, the peregrine is a powerful streamlined bird, a pursuit hunter that preys mainly on pigeons and waterfowl, and is one of the world's fastest birds, able to reach speeds of up to 124 miles per hour (200 kmh). It is slate blue above, with creamy underparts with dark spots and bars, and it has a distinctive extension of dark feathers through the eye like a large teardrop. It has a worldwide distribution, except Antarctica and New Zealand, and across this vast range 22 subspecies are recognized. The peregrine nests on cliff ledges and on tall buildings. As with practically all the diurnal birds of prey, captive reproductive success was not achieved until the late 1960s, when the drastic decline in wild hawks and falcons from the effects of organochlorine pesticides led to increased interest in the raptors and their plight. In the aviaries of institutions such as Cornell University, the Canadian Wildlife Service, and the U.S. Department of the Interior's Patuxent Wildlife Research Center, hundreds of peregrines were produced for reintroduction. The introduction of protective legislation then led to the increased breeding of birds for falconry, and between 1970 and 1975 the number of captive-bred falcons doubled, with half of those produced by private falconers. Within a few years breeders, mostly in the United States, were producing hundreds of falcons annually. Young are now produced regularly by many breeders worldwide, and the peregrine has been hybridized with the merlin and with the prairie, gyr, lagger, saker, and lanner falcons, to produce improved birds for falconry. Falcons need complete peace and quiet to breed successfully, and they are now kept in enclosed aviaries and are viewed through a one-way glass, so they are undisturbed and rarely see humans. They are provided with a sheltered nesting ledge and are fed through hatches so they cannot see their keeper. After a week of natural incubation, some breeders take the eggs for artificial incubation.

American Kestrel   *The most domesticated bird of prey, the tiny American kestrel was the first to benefit from the intense institutional research and breeding programs instigated by the decline of birds of prey due to organochlorine pesticides. Thousands were bred, but private interest in their production has since waned in favor of the larger and more valuable species preferred for falconry.*
Photo: Wesley Aston, Shutterstock.com

American Kestrel (*Falco sparverius*)

The American kestrel is North America's smallest falcon, and was until recently known as the sparrowhawk in the United States. This was confusing because the kestrel and sparrowhawk are completely different species in Europe. It has blue-gray wing coverts that are spotted with white, a rufus back and tail, with a wide black subterminal band across the tail and a white tip on each feather. Its head is white with a bluish-gray top, and there are two narrow vertical black markings on each side of the face. It natural range is from Alaska south to the southern end of South America, and most breeding birds in Canada and the northern United States migrate south for the winter. Kestrels have several different methods of hunting. They may hover, then drop lower and hover again, before dropping straight down onto a rodent in the grass. They may soar in circles looking for prey, and they also hunt from an elevated perch watching for prey on the ground. They also employ direct pursuit, and they catch bats at roost and in flight. Their diet consists mainly of voles, mice, grasshoppers, lizards, and small birds. They are cavity nesters, laying four to six eggs directly onto the rotting wood, soil or sawdust, in the base of tree holes, cliff crevices or artificial nest boxes. The eggs are incubated for 30 days, and the chicks fly at the age of four weeks.

Breeding success with kestrels was not achieved until the late 1960s when the effects of organochlorine pesticides on birds of prey resulted in the establishment of captive colonies, and the kestrel was the first species to be involved in large-scale breeding and research. At the Interior Department's Patuxent Wildlife Research Center, hundreds of kestrels were raised for research on eggshell thinning and other pesticide-related problems. At McGill University's Macdonald Raptor Research Center, manipulation of the photoperiod induced them to breed out of season—in the winter between consecutive spring breeding seasons. Kestrels bred there by artificial insemination in 1974, mated and raised young in 1975, which was believed to be the first time this had occurred in raptors. The semen was collected regularly during the breeding season by the manipulative method used in poultry and through the use of inbred males, which willingly copulated with a collecting device and even mated with a hat worn by their keeper. This increased fertility by 15 percent over the natural matings being recorded by other captive kestrels. Kestrels now breed almost with the ease of chickens, and have hybridized with the prairie falcon; but interest in their production has waned in favor of the larger, more romantic, and certainly more valuable species used by falconers.

6 Mammals

The modern-day domestication of mammals, excluding those in zoos, stems from their use in three completely different ways—as pets, in the laboratory, or on the farm. All three forms of use include some surprising species in some unusual situations, but are a sign of the times—that no land mammal is really beyond the reach of the human desire to utilize it. For most farmed animals and many laboratory ones, the involvement is terminal, in advance of their normal life expectancy, as they are killed for their products or in the name of research.

The term "companion animal" has taken on a whole new meaning. The company of a cat or dog, a white mouse or a guinea pig, no longer satisfies many people, and a host of new mammals now share the home. Similarly, the laboratory community has discovered many new species useful for research. Several of these new animals in the house and the laboratory are already well and truly domesticated, mainly because they are small. The speed of domestication is influenced by an animal's reproductive rate, and in this respect several small mammals have the advantage over all other vertebrates. Four species, the golden hamster, dwarf hamster, chinchilla, and the Mongolian gerbil, are classic examples of how fast domestication can occur. Their early sexual maturity, short gestation periods, and large litters have hastened the process, which has also been aided by selective breeding and the production of many mutants. In just a few decades all four rodents have developed from animals known only to biologists to very familiar house pets.

Several of these new pet species are also prominent in biomedical research. They include the golden hamster, gerbil, chinchilla, degu and ferret in which the pet trade predominates and, in the case of the chinchilla, the fur farm industry also; and the naked mole rat and multimammate mouse, which are kept as pets but are presently both more prevalent in the laboratory than in the home. Another small mammal, a marsupial called the gray short-tailed opossum, has been maintained in research laboratories for almost 30 years, and is also considered domesticated.

The other major aspect of mammal use and domestication in the twentieth century resulted from the discovery that numerous species could be "farmed," and the traditional view of farming therefore needs revising. Deer and antelope now graze the fields in place of cattle and sheep in several countries, and fur-bearers are farmed in a more intensive manner that provokes severe criticism of their lack of space and mode of slaughter.

■ ALTERNATE PETS

In the latter half of the last century, several species of small rodents joined the rat, mouse, and guinea pig as new pet trade species. The first of these was the golden hamster (*Mesocricetus auratus*), also called the Syrian hamster, now a very familiar pet animal; while the Mongolian gerbil (*Meriones unguiculatus*) and dwarf hamster (*Phodopus campbelli*) are rapidly increasing in popularity. The golden hamster is so popular and familiar it may not seem like a new domesticate, yet its captive history dates back only to 1930. It was then that a mother and her large litter were caught in the Syrian Desert by Professor Israel Aharoni of Jerusalem's Hebrew University. Only the golden-colored mother and three babies survived the journey back to the university laboratory, but from those few founders all the world's pet golden hamsters are descended. Hamsters develop rapidly, and females can conceive when just one month old; their gestation period is only 15 days. From Jerusalem some hamsters were taken to England in 1931 and reached the United States in 1938. They were originally considered laboratory animals but soon became popular pets, which is rather surprising considering their solitary nature and unsociable behavior. They now occur in a great variety of colors, and mutants for coat length and size have also been perpetuated, the most popular being the long-haired "teddy bear" hamster.

Several other hamsters have since also become popular pets. The dwarf hamsters (*Phodopus*) are charming, sociable, little creatures; much more agreeable than the golden hamsters, which they will undoubtedly soon usurp as the most popular small mammal pet. Three species are now being bred, the Siberian dwarf hamster (*P. sungorus*), Roborovski's dwarf hamster (*P. roborovskii*), and the most commonly available one—Campbell's or the Dzungarian dwarf hamster (*P. campbelli*) from Mongolia. There is much confusion over the identification of these similar species, so hybridization can be expected. Their intense breeding has already produced numerous color mutants, and in barely three decades this small rodent has experienced a faster rate of domestication than the golden hamster, and therefore probably the fastest rate of all mammals. Another factor very much in its favor is that it does not smell "mousy" even when its cage is infrequently cleaned. Another denizen of Mongolia, the gerbil or jird (*Meriones unguiculatus*), has also become a popular pet in the Western world, beginning in the 1950s. Like the golden hamster, it was originally considered a more suitable subject for the laboratory, but its potential as a pet was soon recognized when it was imported into the United States. A close relative of the gerbil, the larger Persian jird (*Meriones persicus*) is now also of interest to pet breeders. The chinchilla (*Chinchilla lanigera*) was only imported in 1923 into the United States for fur farming, and is a multipurpose animal, kept as a pet, farmed for its pelt, and used extensively in the laboratory. The human control of all these

Dwarf Hamster *The perfect pet rodent, the dwarf hamster is a charming, soft-furred little creature, highly sociable (unlike the golden hamster) and lacking a "mousy" smell. Although only produced by pet breeders in the last three decades, inbreeding and selection has already resulted in several color and pattern morphs.*
Photo: Carolina, Shutterstock.com

rodents resulted in a complete change in their lifestyle, and they have become so easily and quickly domesticated due to their willing acceptance of total confinement, completely artificial diets, plus their inability to partake in any form of social or territorial behavior or mate selection.

Other small rodents now appearing with more frequency in the pet trade are the degu (*Octodon degus*), which is also used in the laboratory, and the giant pouched rat (*Cricetomys gambianus*). Despite the pouched rat's size and its very rat-like appearance, it is an intelligent, friendly, and inquisitive animal, and when acquired as a baby makes a pleasing companion. Their name refers to their cheek-pouches, not kangaroo-style baby pouches, in which they collect and carry food for storing. They are omnivorous and captive animals have thrived on nuts, fruit, dog biscuits, and sunflower seeds. They breed continually throughout the year and are quite social, several animals living together amicably.

Several species of mice are now also available as pets, including the exotic Egyptian spiny mouse (*Acomys cahirinus*), and the striped grass mouse (*Lemniscomys striatus*), plus two native North American species, the white-footed mouse (*Peromyscus leucopus*) and the deer mouse (*P. maniculatus*). The striped grass mouse is the size of a house mouse and becomes quite tame if handled regularly. It has coarse dull-brown fur enlivened with six rows of white spots running from the shoulders to the rump. A native of Africa, it lives in the dry woodlands and grasslands surrounding the great central forests from Sierra Leone to Malawi, where it makes runways through the grass leading to its small round nest. It is a very fecund species, with a gestation period of 21 days and litters averaging six young. The bulk of its diet consists of seeds and grass shoots, varied with locusts. The multimammate mouse (*Mastomys natalensis*), a small dark-brown animal from sub-Saharan Africa, has also been maintained sufficiently long for several color mutants to be produced, including a golden-yellow one with a white belly. It is also a favored research species.

Lemmings, especially the Norway or brown lemming (*Lemmus lemmus*) and the steppe lemming (*Lagurus lagurus*) are now being kept as pets. The Norway lemming is the most colorful species, a golden-brown animal in which the top of the head and nape are black and the underparts creamy white, and it has orange-brown stripes from its eyes to behind the ears. It has dense fur, small fur-covered ears, and a head and body length averaging 5 inches (12.5 cm). Lemmings are prolific breeders, with even greater fecundity than the house mouse, and they are easy to keep, despite original fallacies regarding their diet—that they must have moss to keep healthy, for example, whereas they have thrived on dog chow and sprouted sunflower seeds. Their only disadvantages seem to be their disagreeable nature and the desire of most individuals to bite when they are handled.

Another very new species in the pet trade is the hedgehog, specifically a hybrid between the white-bellied hedgehog (*Atelerix albiventris*) and the Algerian hedgehog (*A. algirus*), which has been given the pet breeder's name of African pygmy hedgehog. It is the most commonly kept hedgehog, although both pure ancestors are also available. Although breeding began only two decades ago, several color phases, including cinnamon, albino, and pinto, have already been produced. The ancient Romans actually kept Algerian hedgehogs in the fourth century BC. They apparently found their quills useful for several purposes, but it is unknown if they bred them, so their domestication may be a recent occurrence. The European hedgehog has been used as a research animal in England, and captive colonies have been maintained for several generations. Hedgehogs are now so well established as pets, and in such large numbers, that both canned and dry commercial foods are available for them.

Although marsupials have been major laboratory animals for some years, especially in Australian universities, several species are now available from exotic pet breeders. They include wallabies, such as the Bennett's, red-necked and tammar wallabies, and the sugar glider—which is now a very popular pet. Its resemblance to the North American flying squirrels, which are rodents, results from convergence—when unrelated animals develop similar characteristics due to living a similar lifestyle. In addition to being produced by caring breeders, hundreds are now also being kept in the equivalent of "puppy mills" in poor conditions.

Carnivores have not been excluded from the recent spate of pet production. One of these is the ferret, as a result of the changing use of a long-domesticated species. Although its original domestication is now obscured by time, it is believed to have been first controlled by man almost two millennia ago. What is certain, however, is that until the last century it was the only carnivore other than the cat and dog to be domesticated. It was used to rid medieval manor houses of their rats, and from the fourteenth century poachers and gamekeepers in England put ferrets down rabbit's burrows to flush them out. It has recently become a popular house pet, is widely used in research laboratories, and is farmed for its pelt in Finland. Unlike its vague early history of domestication, the ferret's introduction into New Zealand as a control of the previously introduced rabbits is well recorded. It has changed little in the centuries it has been domesticated, and is still being produced in both the basic creamy-yellow mutant, and the "polecat" type that resembles its wild ancestor. About three decades ago in Norway, however, a long-haired mutant appeared in a litter, and "angora" ferrets are now produced. Commercial diets are now available for ferrets.

Many new cats have also been produced in recent years. A number of these resulted from the selective breeding of house cats, and the production of new breeds such as the ocicat, a spotted cat developed from cross-breeding Abyssinian and Siamese cats; and the Bombay cat, a short-haired shiny black animal with coppery-colored eyes. However, these are just new breeds developed from a long-domesticated animal, and have no relevance to recent domestication. Most of the new cats produced in the last century, however, are the products of crossing wild cats with house cats, plus the hybridizing of several species of wild cats and the resultant loss of pure blood, which is unfortunate. Also, purebred wild cats are now being bred for the pet market.

The ancestor of the house cat (*Felis catus*) is the African wild cat (*F. libyca*), with possibly genes of the jungle cat (*F. chaus*) added more recently, as sailing vessels plying between India and Britain during the eighteenth century commonly carried jungle cats aboard as ratters, and some are believed to have disembarked at British ports. There is now a thriving business involving the production of new cats through the interspecific hybridization of the house cat with a number of small wild cats. The production of these hybrids is primarily for commercial purposes. Many species, including Geoffroy's cat (*Oncifelis geoffroyi*), the little spotted cat (*Leopardus tigrinus*), the margay (*Leopardus weidi*), and ocelot (*Leopardus pardalis*), have 36 chromosomes, whereas the house cat has 38. Offspring from crosses between the two generally have 37 chromosomes, 19 from the domestic parent and 18 from the wild cat. In most of the wild and house cat hybrids the female offspring are fertile, and male offspring are occasionally fertile. The offspring are neither true breeds (as only one parent was domesticated), nor are they species (as only one parent was a wild animal), and are currently called "designer breeds."

In the United States, female jungle cats (*F. chaus*) have been mated with male house cats to produce the breed called the chausie. This hybrid has then been crossed with the European wild cat (*Felis silvestris*) to produce the Euro-chausie. The house cat has also crossed interspecifically with serval (*Leptailurus serval*), margay, caracal (*Caracal caracal*), Amur leopard cat (*Felis bengalensis euptilura*), little spotted cat, black footed cat (*F. nigripes*), the bobcat (*Lynx rufus*), and the European wild cat. Also, house cats have mated with the African wild cat (*Felis libyca*), their ancestor, so their offspring are interspecific hybrids, as the house cat is now considered a separate species to its wild ancestor. Crossing the house cat with the Bengal leopard cat (*F. bengalensis*) has produced the very attractive Bengal cat, and hybrids between Geoffroy's cat and the house cat, called safari cats, were first produced in the 1970s. The savannah cat is the offspring of a house cat and serval. The desert lynx was produced by crossing the house cat with the bobcat, and is now being bred in a variety of colors, including ebony, silver, snow, chocolate, and patterns such as spotted, marble, and tawny. Bobcats and feral house cats have also mated and produced offspring in the wild.

In the drive to produce new alternate, and expensive, exotic animals for the pet trade, pure species of wild cats have been hybridized. The "caraval" is a cross between a male caracal and a female serval, two African grassland species, the offspring having the caracal's tawny coat covered with the serval's dark spots. The "servical" results from breeding a male serval and a female caracal, and is tawny with pale spots. Canada lynx (*Lynx canadensis*) and bobcat, a cross that has occurred naturally, are also hybridized. The bobcat and the European wild cat have been crossed with the jungle cat, and a female ocelot mated to a male margay has produced offspring. The hybridizing of species in this manner harkens back to the days when zoos deliberately crossed large cats, especially lions and tigers—fortunately a practice largely abandoned.

Several species of purebred wild cats are also now available as pets, including the serval, caracal, Siberian lynx, Canada lynx, bobcat, and ocelot. Breeders generally hand-rear the kittens to ensure they are tractable and accustomed to people and to the home, and some extol the conservation value of breeding such animals. It is claimed that breeding for the pet trade reduces the drain on wild populations, but in reality there is no longer any legal pet trade of wild-caught cats, as all species are listed in either Appendix I or II of CITES. If these animals were not domesticated as a result of zoo breeding, which was the origin of most of the founder stock of the pet breeder, they certainly are now.

Domestic dogs (which are normally just called dogs as opposed to the wild dogs or canids), dingoes, and their common ancestor the wolf, all have 39 pairs of chromosomes, and are therefore able to hybridize. This also includes the coyote and the jackals, excluding the golden jackal. Consequently, wolves have been crossed with dogs, especially similar ones such as German shepherds and sled dogs. Dogs have also been crossed with coyotes and jackals, and it is believed that very few pure "wild" dingoes remain in Australia due to mating with dogs. The dingo is considered a feral domesticated dog, its ancestors having been introduced into Australia by the aborigines, and it hybridizes freely with dogs. So there are now dingoes, feral dogs, and dingo/dog hybrids in the Outback.

The intentional hybridization of captive wild dogs and dogs, however, has not been successful, because the wild parent's genes have been dominant, and the offspring have generally been so difficult to handle that breeders did not pursue their experimental matings. The pups from matings between coyotes and dogs, for example, known as coydogs, inherit their wild parent's dominant solitary behavior and are very antisocial creatures. The other wild dogs—the foxes, raccoon dog, and bush dog—are less closely related and do not breed with dogs. Consequently, there have been far fewer opportunities to create new breeds through hybridizing wild dogs and dogs than there has been with cats. One notable exception is the Sulimov dog, which is 75 percent husky and 25 percent jackal. It has an exceptional sense of smell and is currently used at Moscow's Sheremetyevo Airport as a bomb sniffer.

Breeding wild dogs regularly is therefore generally the prerogative of zoos and wildlife parks, where domestication is occurring, but not with the intent of starting new breeds, as they are normally kept pure. Apart from the domesticated fur-farmed canids—the red fox (*Vulpes vulpes*), Arctic fox (*Alopex lagopus*), and raccoon dog (*Nyctereutes procyonoides*)—the only other domesticated species is the fennec fox (*Fennecus zerda*), the only wild dog now bred regularly for the pet trade, although hand-raised Arctic foxes are also kept as pets. It is likely, however, that other small wild dogs will eventually be of interest as pets. This would include such species as the bush dog (*Speothos venaticus*), when it has recovered from its current threatened status; the raccoon dog (*Nyctereutes procyonoides*); the kit fox (*Vulpes velox*); the swift fox (*V. macrotis*); and the gray fox (*Urocyon cinereoargenteus*).

While most aspects of new mammal production are relatively harmless, except the hybridizing of wild cats, the breeding of monkeys specifically for the pet trade causes considerable concern. The difficulty of legally acquiring wild-caught monkeys as pets, which was once so very easy to do, has unfortunately not totally eradicated the pet monkey business, as primates in private hands are now being bred specifically for sale as pets. The small primates now being produced for the pet market include the white-faced capuchin (*Cebus capucinus*), vervet monkey (*Cercopithecus aethiops*), spider monkeys (*Ateles*), ruffed lemur (*Varecia variegata*), squirrel monkey (*Saimiri sciureus*), and the long-tailed macaque (*Macaca fascicularis*), which have all been offered for sale recently. Monkeys have been hybridized by pet breeders. The tufted capuchin (*C. apella*) has been crossed with the weeper capuchin (*C. olivaceous*), and the rhesus monkey (*M. mulatta*) has been mated with the red-faced macaque (*M. artoides*). Naturally these house monkeys are very humanized, bottle-fed, raised in diapers, and accustomed to waist belts, as monkeys must not be tethered with neck collars. With their articles on animal care, pet magazines have encouraged the "reduction" or "alteration" of the monkey's teeth, both euphemisms for extractions, for their owner's safety. It is obviously totally unnatural to incorporate a gregarious primate into the social fabric of the human family, and to avoid social conflict it has been recommended they are kept in female/female pairs and male/male pairs, which also has the advantage of eliminating breeding and market competition.

■ SOME OF THE SPECIES

Mongolian Gerbil (*Meriones unguiculatus*)

This is one of the most recently domesticated pet and research rodents, with a similar history to the golden hamster. It is a native of the dry grasslands and semideserts of Mongolia, southern Siberia, and northern China, and its captive history began when a few animals were caught in eastern Mongolia in the 1930s. Some were exported to the United States in 1954, and the first animals arrived in England in 1964. Initially, like the hamster, they were considered laboratory subjects, but their potential and value as pets soon became apparent. Also known as Mongolian jirds,

Squirrel Monkey *Although legislation, for both conservation and health reasons, has effectively ended the importation of wild monkeys as pets, several species, like the squirrel monkey, are now produced by breeders as humanized, companion animals, an unnatural and highly criticized practice. They are usually bottle-raised from an early age to ensure their tameness.*
Photo: MaleWitch, Shutterstock.com

the gerbil's fecundity made them ideal subjects for both pet breeders and self-sustaining laboratory colonies. They are sexually mature at 70 days, can breed year-round, and produce up to 12 young after a gestation period of just 21 days, and the babies are weaned when they are 25 days old. They are currently widely used in biomedical studies in parasitology, endocrinology, and neurology, as well as in cancer research. The pet industry, with its intense interest in new mutants, has already produced many color forms through inbreeding and selective breeding.

The wild gerbil is dark gingery brown with pale underparts, is about 5 inches (12.5 cm) long, and has a furred tail about 4 inches (10 cm) in length. Color mutants currently available include solid-colored animals, such as black, slate blue, lilac—which is a light gray—and the ruby-eyed red, which has a white coat. Others are the argente-nutmeg, which is reddish gold, and the Himalayan, which, like the rabbit of the same name, is white with dark ears and nose. When they have two or more colors, one of these being their whitish belly, they are called non-self-colored gerbils. Mutants of this kind currently being bred include agouti, which is reddish brown; champagne, which is off white; the golden-colored, dark-eyed honey morph; and the pied cinnamon, which is white with orange patches. The only hair mutant to appear so far has been the hairless gerbil, which was fortunately not perpetuated.

Egyptian Spiny Mouse (*Acomys cahirinus*)

Despite its name, and its alternate of Cairo spiny mouse, this species occurs throughout North Africa and Asia Minor, and also on the island of Cyprus. Several related spiny mice extend this range across central Asia and southward to the tip of Africa. It is often confused with the spiny rat (*Proechimys*) of South America, a larger and less frequently available animal. Spiny mice are mouse-sized, weigh only 3.2 ounces (90 g) when adult, and are about 4 inches (10 cm) long. Their long, naked, and scaly tails are very brittle and break off easily, but unlike lizards that can regenerate their lost tails the spiny mice must remain tailless. They have large naked ears and very large, black eyes and vary from a rich golden tan to pale yellow and reddish brown, but always with white underparts, except for melanistic individuals. But their most characteristic feature is the thick and stiff hairs on their backs, resembling spines or quills. Spiny mice are animals of the arid and semiarid regions, and although mainly terrestrial in their habits, they can climb trees. They are becoming increasingly popular as pets as their urine is odorless, and they are very docile and friendly animals.

In their natural habitat, spiny mice are nocturnal and crepuscular and are rarely seen during daylight, but they change their habits readily when kept as pets. They are easy to keep and to breed, which they do frequently as there is a post-partum (after birth) estrus and immediate mating, with 12 consecutive litters being recorded during one female's three-year life span. Their gestation period is 35–40 days, which is rather long for a mouse, so the young are quite well developed at birth and their eyes open soon afterward. Consequently they are weaned when only two weeks old, three weeks before their mother gives birth again, and they are sexually mature when seven weeks old. Spiny mice are highly omnivorous and eat seeds, dates, and insects, and in their homelands have raided tombs to eat dried flesh and bone marrow. They are gregarious animals that live happily in groups, the females with young assisting new mothers by biting off the umbilical cord, for example, and helping with the cleanup. Spiny mice have been kept as pets since the 1980s, and several mutants are now available. They have also been used in the laboratory since the 1960s, for research into diabetes and obesity.

Degu (*Octodon degus*)

Degus are common rodents from central Chile that have become popular pets and are also bred on a large scale for the laboratory. They weigh about 9 ounces

Degu　*One of the most recent alternate rodent pets, degus have also been widely used in the laboratory. They become very tame and are soon accustomed to being handled. Unlike most rodents, baby degus are furred at birth and are soon mobile.*
Photo: Yuriy Maksymenko, Shutterstock.com

(255 g) when adult, are about 7 inches (18 cm) long, and have a tail almost the same length, which is furred and has a black tuft at the tip. The typical wild degu has soft brown fur with creamy colored underparts, but in its short history as a recent pet, black, white, and gray mutants have already been produced. Degus are diurnal and are burrowers, and in their native land they store food in their burrows for the cold winter months. They are very social animals, and several females may raise their young in the same burrow; and when they are kept in pairs, the father is attentive at the birth and then helps to care for the young. They are sexually mature at four months and breed throughout the year. Their gestation period is about 90 days, very long for such a small rodent, and they have up to 10 young per litter, which are weaned when they are about five weeks old. As a result of their long gestation period, and unlike most rodents, at least the smaller species, baby degus are precocial at birth. They are furred and have white teeth (which later change to the typical rodent orange), their eyes are open, and they run around within a few days. Degus become very tame and accustomed to being handled, but should never be picked up by the tail, as they may spin to escape and shed their tail skin, exposing the flesh and bone which the animal will then amputate. They are one of the more recent rodent pets, popular for only a few years; but already the market is considered overloaded, with breeders producing just too many animals for the current demand. They are susceptible to diabetes and have therefore been useful as laboratory animals for much longer than their history as pets.

Sugar Glider (*Petaurus breviceps*)

Although it resembles a North American flying squirrel (which is a rodent), the sugar glider is actually an Australian marsupial, with a pouch in which it raises its babies in the manner of the kangaroos. It is about 7 inches (18 cm) long, has a furred tail of about the same length, and weighs 4 ounces (112 g). Silvery gray in color with black ears and a creamy belly, it has a dark stripe through the eye to the ear and another from its nose along the spine to the base of its tail, which is gray with a black tip. Its most distinguishing feature are the patagia, flaps of fur-covered skin stretching down the sides of its body from its hands to its feet. When it leaps from a tree and stretches out its limbs and the flaps, it can glide downward for a distance of up to 150 feet (45 m), turning its hind feet up at the last moment to alight in a vertical position on all fours on the destination tree trunk. Any resemblance to a flying squirrel is purely coincidental and results from totally unrelated animals living a similar lifestyle, even on different continents.

Like the flying squirrel, the sugar glider is nocturnal and sleeps in a tree hollow during the day. When abroad at night, its enemies are owls and the native cats or dasyures—marsupials that have developed similar predatory habits to the small cats of other lands. Several subspecies of gliders are recognized within a range that encompasses eastern and northern Australia and New Guinea and its neighboring islands, and they have been introduced into Tasmania. They are protected and cannot be kept as pets in Australia, nor can they be legally exported. But they have been acquired by breeders in the United States, and many have also been legally exported from Indonesia, so most of the current North American population are probably subspecific hybrids. However, keeping them as pets is illegal in some states, notably California, Georgia, Hawaii, and surprisingly Alaska, due to the possibility of escapees becoming established and being destructive to native plants or animals. Sugar gliders make interesting, active (at night), and inquisitive pets, but they are social animals and need a lot of attention, preferably provided by their own kind. Their captive life span may be 15 years, so keeping them is a long-term commitment.

Mutants, especially leucistic and albino individuals, have appeared in the wild occasionally, and as with practically all mammals that are continually bred, and probably inbred, mutant babies sooner or later appear in captive glider's pouches. Consequently, several morphs are already being offered by breeders. They include leucistic gliders—white animals with black eyes, and therefore not albinos; the red-cinnamon, with a red coat and reddish-brown markings; and the chocolate-brown glider, which has a brown coat with darker brown markings. There are now many breeders of gliders in the United States, including several that have been termed "mills" where hundreds of gliders are kept in unsatisfactory conditions, akin to the production of dogs in puppy mills.

Fennec Fox (*Vulpes zerda*)

The fennec fox is the smallest member of the wild dog family *Canidae,* with a head and body length of 16 inches (40 cm), a 10-inch-long (25 cm) tail, and a weight of just over 3 pounds (1.3 kg). It is the most charming of all the fox-like dogs, with a soft, sand-colored coat and a bushy, black-tipped tail. Fennec foxes breed once annually, in the spring, and their gestation period of 51 days is about 12 days less than the domestic dog. Litter size varies from one to five kits, which have a body covering of light sandy-colored fuzz. Their ears are folded over and their eyes are closed at birth, but they open at the age of 10 days, and their ears stand up soon afterward. Breeders usually remove the kits from their mother for hand-raising soon after birth, as tame and handleable babies make better pets and are therefore more valuable. Fennec foxes are usually given the run of the house, but care must be taken not to allow them outside or they will likely never be seen again. They have been trained to the harness and leash and to use a litter box.

Hearing is the fennec fox's most highly developed sense, and it has the largest ears in proportion to its size of all the wild dogs, able to pick up the smallest sounds made by its prey. It has very large bullae or earbones that also improve its hearing. The possession of a tapetum lucidum, the reflective layer behind the retina, increases its night vision. The fennec fox's natural habitat is the desert zones of North Africa and the Middle East, a region of extreme summer heat, and its large ears also help to radiate heat from its body. It gets sufficient fluids from its food to manage without free water for long periods, and spends the hot daylight hours under the shelter of rocks or in its tunnels, which may be several yards long. It appears at dusk to hunt gerbils, jerboas, insects, and lizards, and its hairy soles enable it to run fast on soft sand when chasing its prey. It is also fond of dates. The fennec fox's major predators are eagle owls and caracals. It breeds readily, and is now being produced commercially as one of the latest "alternate" house pets. It is a such a delightful little dog there is no doubt that it will eventually be totally domesticated, and a breeder's registry has already been established to help avoid the dangers of inbreeding.

Bengal Cat (*Felis catus*)

The Bengal cat is the product of considerable selective breeding over many generations, originally involving domestic house cats and the Asian or Bengal leopard cat (*Felis bengalensis*). The current animals contain genes of a number of house cat breeds, including the American short hair, the Abyssinian and the ocicat, together with those of the wild leopard cat and leopard cat/house cat hybrids. The result is a very exotic-looking animal with the house cat's temperament. It is therefore not exactly a breed as it is part wild cat, nor is it a species as it is part house cat. The naming problem has been solved by calling it a designer breed. The Bengal cat is a striking and alert animal, very active, playful, and intelligent. It has already been bred in a wide range of colors and patterns, including brown, orange, charcoal, and ivory, with rosettes or spots, sometimes with a darker outline, on the body, and with barred legs and tail. An unusual pattern mutant is the marbled Bengal cat, derived from the genes of the leopard cat and a domestic tabby. Exhibition standards for Bengal cats have been set by The International Bengal Cat Society, and there are currently three new varieties of cats being developed. They are the Serengeti cat, resulting from the crossing of a Bengal cat with a short-haired breed such as the Siamese, the offspring resembling a serval. The toyger cat is another cross between a Bengal and a short-haired house cat to produce a striped cat that resembles a tiny tiger. The cheetoh cat, with finely spotted body and striped legs, is the product of crossing the Bengal with another domesticated spotted cat (such as the ocicat), with the intention of producing a wild-looking cat with a gentle disposition.

■ LIFE IN THE LAB

Rodents still have pride of place in laboratories, mainly because of their small size and rapid reproductive rate—their early sexual maturity, short gestation period, and high birth rate. But although the traditional domesticated white rats and white mice are dominant in numbers, the list of research laboratory rodents now includes a host of others. It includes several species of hamsters, now used in biomedical and medical research, with at least one million used annually in research laboratories in the United States. The most plentiful one is the golden or Syrian hamster, the third-most common laboratory animal after rats and mice, and a research species since soon after its capture and breeding in 1930. It is used in virology and parasitology research; its cheek pouches are highly regarded for studies on the transplantation of tissues, and its teeth for studying periodontal disease. The Chinese hamster (*Cricetulus griseus*) has been a laboratory animal since 1949, its susceptibility to diabetes mellitus making it a useful research animal. Despite its domestication, however, the production of inbred strains—which is accepted as 20 generations of single linebreeding—took almost 40 years to achieve. Armenian hamsters (*Cricetulus migratorious*) have been laboratory animals since 1963, and with the Romanian hamster (*Mesocricetus newtoni*) and Turkish hamster (*M. brandti*) are used in cytogenetic studies—the origin, variation, and development of cells. The guinea pig–sized European hamster (*Cricetus cricetus*), a very unsociable animal, is used for research into carcinogenesis. None of these animals are imported from the wild, so laboratory populations are descended from many generations of captive-bred, domesticated individuals.

The Mongolian gerbil is the second-most popular rodent to be domesticated during the last century, soon after the golden hamster. Wild-caught gerbils were first used in the laboratory in 1933, when their susceptibility to the blood-fluke (*Schistosoma*) was discovered. Their potential value then resulted in the capture of 20 pairs in the Amur region of far-eastern Asia in 1935. They were taken to Japan and maintained in a closed colony—without the addition of further importations. A few pairs were imported into the United States in 1954, and from there offspring were sent to Great Britain and Europe. These animals resulted in a very large captive population of pets and many laboratory strains.

The multimammate mouse (*Mastomys natalensis*), one of Africa's most common rodents and a new pet species, has been a laboratory animal since 1939, and on one occasion was intensively inbred for 20 generations. It has great value for research into stomach cancer and spontaneous tumors, and it has the highest rodent susceptibility to osteoarthritis. It is also the only known nonhuman host of the Lassa virus, which kills 5,000 people annually in West Africa. The carrier mice do not become ill, but are infectious when they shed the virus in their feces and urine. Other rodents now prominent in biomedical research are the degu for studies into congenital cataracts and diabetes, the Prairie deer mouse (*Peromyscus maniculatus bairdii*) as a model for sleeping sickness research and for idiopathic epilepsy, and lemmings for skin cancer research. Naked mole rats (*Heterocephalus glaber*) have lived in self-sustaining laboratory colonies for many generations and have become models for studies into aging. This makes sense as they are the longest-lived small rodents, often reaching the age of 20 years and occasionally longer.

The most surprising animals to find in the laboratory are marsupials. They could perhaps be expected in Australia, the marsupial's domain, but they are also used in several other countries. In Australia the tammar wallaby (*Macropus eugeii*) is maintained in self-supporting colonies at several universities, and in one—Melbourne's LaTrobe University—a self-sustaining colony dates from 1992. It also has a closed colony of the Tasmanian long-nosed potoroo (*Potorous tridactylus apicalis*). The fattailed dunnart (*Sminthopsis crassicaudatus*) is also widely kept; the laboratory colony at the University of Adelaide was founded in 1978 and at one time (1995) comprised 1,000 animals. In the United States, the Virginian opossum is a laboratory animal involved in immunology and burn research. But the most widely and continually bred marsupial, and therefore the most domesticated species, is the gray short-tailed opossum (*Monodelphis domestica*), a tropical American animal that has been intensively captive-bred since 1979, and 70,000 pedigreed animals have been produced.

Shrews, considered for many years the most difficult animals to keep alive, have been maintained and bred in the laboratory for many years. At Tel Aviv University, pairs of white-toothed shrews (*Crossidura russula*) were kept in pairs in small stainless steel containers with wire mesh covers, similar to laboratory rodent cages, and were bred for many generations. The nine-banded armadillo (*Dasypus novemcinctus*) is kept as a laboratory animal because, like humans, it is susceptible to leprosy; but develops antibodies to protect itself. The laboratory armadillos are therefore artificially infected with leprosy as part of the process to produce vaccines for humans.

Two carnivores other than cats and dogs have also been used for biomedical research. Both members of the family *Mustelidae*—which includes the weasels, badgers, and otters—they are the mink, recently domesticated on fur farms but also a laboratory animal; and the ferret, domesticated descendant of the polecat. The mink is used as a model for human viral infections and for Creutzfeldt Jacob disease, an invariably fatal brain disorder, and its variant—bovine spongiform encephalopathy (BSE), commonly known as "mad cow disease." The most widely used carnivore, however, is the ferret. The ultimate all-purpose animal, it was used many years ago as a rodent control, then for rabbiting. It has been a laboratory animal since the 1930s, especially useful for microbiology due to its susceptibility to viral infections—notably canine distemper and human influenza. More recently, it has become a popular house pet.

Due to their close neurological, immunological, and reproductive relationship to humans, primates have for many years been favored animals for biomedical research, and their use is well known and frequently criticized. Until quite recently they were subjected to such massive exportations from their native lands that some regions were virtually depopulated. In East Africa the predator-prey relationship was affected, the dearth of baboons causing leopards to seek other prey, often livestock, to their detriment. Thousands of monkeys, mainly rhesus macaques (*Macaca mulatta*) and long-tailed or crab-eating macaques (*M. fascicularis*), known also as cynomolgus monkeys, were imported for polio vaccine production in the 1950s and 1960s. When the supply of these primates began to dwindle, emphasis moved

Baby Ferrets   *Ferrets are the most adaptable and multipurpose domesticated carnivores. They were first used by man in medieval times to control vermin, then by poachers for rabbiting, and in the last century they became important laboratory animals. They are farmed in Finland for their pelts, and are now also very popular house pets.*
Photo: Gila R. Todd, Shutterstock.com

to other species, particularly the vervet monkey (*Cercopithecus aethiops*) and the squirrel monkey (*Saimiri sciureus*). Most of these animals were involved in terminal research, so domestication was not an issue.

With the virtual cessation of wild-caught imports, however, the emphasis in research establishments changed to providing their own stock, and many captive-breeding colonies were established, with the advantage of producing healthy off-spring of known parentage and medical history. In the United States, the National Primate Research Centers (NPRC) house large collections of nonhuman primates, many of them in breeding colonies, where natural mate selection within the colony can occur and some foraging is possible. Life in the lab does not necessarily mean close confinement, nor the infliction of pain.

Several of the eight NPRC facilities maintain large colonies of rhesus macaques, crab-eating macaques, and pig-tailed macaques (*Macaca nemestrina*). The Oregon NPRC keeps 70 percent of its primates in social units in large outdoor enclosures. Other species maintained and bred by the centers include squirrel monkeys, the dusky titi (*Callicebus moloch*), the brown capuchin (*Cebus apella*), common marmoset (*Callithrix jacchus*), and several species of baboons (*Papio*). The herbivorous baboon's circulatory system handles fats in a manner similar to man, and like man they develop arterial lesions, so they have been of great value for studying the relationship of diet to human cardiovascular disease.

At the Wisconsin NPRC, the common marmoset colony produces 75 offspring annually, and other primates maintained as breeding colonies include the 300 owl monkeys (*Aotes trivirgatus*) and 450 squirrel monkeys at the University of South Alabama. The Semel Institute at the University of California at Los Angeles has a research colony of vervet monkeys that produces 100 babies annually. Chimpanzees are bred at the University of Louisiana at Lafayette, where the colony contains 360 animals. At Cayo Santiago, off Puerto Rico, a free-living colony of rhesus macaques has lived in seminatural conditions for many years. A semifree-ranging colony of Japanese macaques, caught as an intact troop at Arashiyama, Japan, has been living since 1972 in a fenced area of 108 acres (43 ha) at Laredo, Texas.

■ SOME OF THE SPECIES

Gray Short-tailed Opossum (*Monodelphis domestica*)

This small marsupial is prominent in captivity, and although it has quite a following as a pet species, it is known mainly for its role in the laboratory. A New World animal, it is a descendent of the marsupials that remained behind when Gondwanaland broke apart long ago and carried the others to Australia. It has been maintained for so long it is now also called the laboratory marsupial. Its specific name *domestica* was actually based, years ago, on its preference for living in human dwellings, where it was encouraged as a control of vermin and insects, but the name is now even more appropriate.

The short-tailed opossum is a mouse-sized, gray-coated animal with a bare, fully prehensile tail, and is a native of eastern and central Brazil, Paraguay, and Bolivia. It is highly suited for a life of complete control by man, being so very adaptable that it can be kept (singly) in cages made for laboratory rats, although it is moved to larger cages for mating. Although naturally quite carnivorous, it has thrived on a diet of laboratory rodent pellets. It also meets the other major requirement for intensive captive breeding—high fecundity. It is very prolific, giving birth to up to 12 young in a litter after just 14 days' gestation, and can produce three litters annually. Like most New World marsupials, the short-tailed opossum lacks a pouch, and at birth her babies immediately attach themselves to a teat. The mother then stays in her nest for most of the next four weeks, just venturing out briefly to feed, while the babies hang on. When they are four weeks old they begin to detach from the teats and then ride on her back when she leaves the nest.

The gray short-tailed opossum is the most common marsupial in research. Since 1979 a large colony has been maintained at the Southwest Foundation for Biomedical Research at San Antonio, one of the world's leading independent biomedical research institutions. The Foundation since that time has produced 70,000 animals whose pedigrees have been fully documented, and it currently provides 2,400 animals annually to establish research colonies at other biomedical centers around the world. The opossums are used as models for skin cancer and hypercholesterolemia research, and for studies into spinal cord injuries, as young animals are able to repair damage to their spinal cords. It is the first marsupial to be sequenced at the National Human Genome Research Institute as a model for marsupial genetics research.

Rhesus Macaque (*Macaca mulatta*)

The rhesus macaque is a medium-sized brown monkey with paler underparts, and a red face and red buttocks when adult. Mature males may have a body length of 24 inches (60 cm), and tails 12 inches (30 cm) long, and can weigh up to 26 pounds (12 kg). Females are just over half the male's size. They are natives of southern Asia from Afghanistan to southern China, encompassing India, Myanmar, and Thailand, and have been introduced into Florida, where they are now established. They live in multi-male, multi-female groups controlled by a matriarch. Very opportunistic animals, rhesus monkeys thrive in all types of habitat, virtually wherever food is available; and in India they frequent towns and villages, where they raid homes and crops, but the significant damage they cause is tolerated by Hindus who consider them sacred. They have been favorite laboratory animals for almost a century and were once the subject of large exportations from their native lands, to the extent that many regions were virtually depopulated of their young stock and future breeding was compromised. Eventually they were protected and exports were banned or more strictly controlled.

The rhesus monkey is considered a model primate for many kinds of research, due to its similarity to man in genetics, physiology, and metabolism, even though it is less closely related to humans than the chimpanzee. It has therefore been widely used in medical and biomedical research. The hereditary antigen Rh-factor, discovered in rhesus monkey's red blood cells in 1940, helped to determine human blood groups; and NASA launched rhesus monkeys into space in the 1950s and 1960s. Its genome is currently being sequenced by the Human Genome Sequencing Center at Baylor University.

The original research relied on a regular supply of replacements from the wild, but the emphasis is now on breeding. Rhesus monkey colonies are maintained by several universities and at some of the National Primate Research Centers. The Oregon NPRC maintains the largest pathogen-free collection of rhesus monkeys in the United States; most of their 3,800 monkeys live in colonies in large outdoor corrals, which were pioneered there in the 1970s. The founders of this group arrived in Oregon in 1964 from Brown University, where they had been for several years. The animals in this and other similar long-established breeding colonies are obviously now in the early stages of domestication. The 950 rhesus monkeys on the 38-acre (15.3 ha) Cayo Santiago island off Puerto Rico, descendants of animals released there in 1938, are less rigidly controlled. They are allowed the run of the island, but are provided with 50 percent of their food, getting the balance from natural foraging. Only researchers and the provisioners are allowed to land on the island.

Chinchilla (*Chinchilla lanigera*)

Chinchillas are small colonial rodents with a head and body length of about 12 inches (30 cm), a fluffy tail about 6 inches (15 cm) long, and large ears. Their natural or wild coloration is gray with paler underparts, and they have longer black hairs on

Rhesus Monkeys *A group of rhesus monkeys forage naturally in their large outdoor enclosure at the Oregon National Primate Research Center, where such large corrals for laboratory primates were pioneered in the 1970s. A model primate for many kinds of research, rhesus monkeys are now increasingly being maintained in large colonies, where natural mate selection and food gathering is possible.*
Photo: Courtesy Oregon National Primate Research Center.

their backs. Males weigh up to 17 ounces (500 g), while the females may weigh 22 ounces (750 g). They are high-altitude animals, crepuscular or nocturnal in their habits. They shelter during daylight among the rocks in burrows or crevices, in the cool, dry, rarified air of the Andes of Peru, Bolivia, Chile, and Argentina, and are named for the Chincha people of the mountains. They are very agile animals, able to jump several feet into the air when escaping their predators, mainly small cats. Chinchillas are most well known for their pelts, one of the world's most coveted furs, of which about 80 are needed to produce a full-length coat. Chinchilla fur is the most luxurious and softest of all land mammals, with the highest density—about 20,000 hairs per square centimeter. Over 50 hairs grow out of a single follicle, unlike the human follicle that produces a single hair, and chinchilla hair is 30 times softer than human hair. The Andeans made garments out of chinchilla pelts long before the arrival of the Spanish conquistadores, but overtrapping by Europeans almost exterminated them.

Mining engineer M. F. Chapman saw the few surviving chinchilla in the Andes in 1918, and was determined to take some back to the United States. Previous attempts had failed, as the animals died when they left the cool and dry atmosphere of the mountains for the heat and humidity of the coastlands. Chapman succeeded in getting capture and export permission, and employed the native Andeans to catch some. After considerable effort they caught 11 chinchillas at an altitude of 12,000 feet (3,657 m), which were brought down to sea level in stages over several months to acclimate them slowly to the lower elevations. During the sea voyage home to the United States, their cages were cooled with ice and wet towels. All 11 animals survived, plus one born at sea; they became the founders of the chinchilla farming industry, and the animal's eventual popularity as pets and in the laboratory. Fur farming and their domestication is therefore credited with saving the chinchilla from extinction.

The selective breeding of chinchillas for the fur trade has produced many mutants including white, ebony, brown, sapphire, and chocolate, plus a white one in which the hairs are tipped with silver. In addition, their use in research has been invaluable. In the 1940s they helped develop cholera vaccine, and more recently have been used in outer space sleep research. As they have similar middle ear and auditory nerve systems to humans, they have been models for hearing studies and children's middle ear infections. Chinchillas are also used in studies relating to aging and high-altitude sickness, and as models for Chagas disease, caused by the parasitic protozoan *Trypanosoma cruzi* that kills 50,000 people annually in tropical America.

■ TERMINAL FARMS

Until quite recently, farming meant raising crops or traditional animals such as cows, pigs, and chickens, but the last century completely changed that perception, and farming now includes deer, antelope, mink, foxes, and even rodents. These animals are kept primarily for their meat, skins, and hides, and by-products such as antlers and the contents of their musk glands. Therefore, most of this modern form of farming is terminal in nature, the major products being available only after the animal's death, which is often at an early age or usually well in advance of its normal life expectancy. Mink are killed for their pelts when they are nine months old, and plains bison have reached an acceptable killing weight when they are three years old. Only the breeding stock live anywhere near their natural life span, at least their natural breeding-life span, for like traditional livestock, farmers of the new animals must dispose of worn-out breeders and retain young from each year to replace them. However, the captive elk in North America, currently about 140,000 strong, are considered too valuable to kill for meat, and are being retained as breeders and for annual antler-cutting until the population is much larger.

With few exceptions, the housing of these new commercial domesticates is appropriate for the species. Deer, bison, and musk oxen graze fields once monopolized by Holsteins, Jerseys, and Herefords. The small fur-bearers and the rodents are kept in wire cages, to which there really is no alternative other than providing more space. The major exception to the rule is the manner in which deer are farmed in China and Taiwan, in small, brick-floored yards enclosed by high walls, that to the Western mind are more appropriate for intensively housing pigs than deer.

Like all the commercially produced, newly domesticated animals, these farmed species are all artificially selected, either purposely or accidentally, for their ability to thrive and breed in the conditions and husbandry provided, and for their tameness and ease of handling. More specific selection is also involved, for larger antlers, more luxuriant fur, or less fat and marbling of the meat. Only in the fur bearers is color important, and numerous mutants of mink, fox, and coypu are bred, making the final products more alluring and extending the buyer's choice. Hybridizing species, such as silver fox and Arctic fox, and red deer and Pere David's deer; and subspecies, like the red deer and elk, also extend the range of commercial possibilities. Commercial selection by deer farmers includes breeding animals that convert food to meat quicker and are therefore marketable at an earlier age, and are consequently more profitable.

This new form of farming has been touted as a great benefit to conservation, but it is primarily a terminal operation that has little or no conservation value. It is often confused with wildlife ranching, which has limited conservation value. Ranching is practiced in several countries, especially in South Africa, where there are 5,000 farms on which free-ranging indigenous antelope, which thrive on the native herbage better than domesticated livestock, are cropped annually. Ranching aids ecotourism, provides for some trophy hunting, and also serves to protect endangered species. Unfortunately farms are often overstocked, causing land degradation as too many game animals can destroy the habitat just as easily as domestic livestock. Human selection also occurs, for trophy hunters obviously target the prime animals, as do those culling the herds for meat.

Although game farming is also mainly a terminal operation, in a few instances, such as the farming of giant cane rats and musk deer, there is hope that it may reduce the pressure on the dwindling wild populations. Some animals have been saved by farming, although this has yet to happen for the musk deer, even after half a century of control. The North China sika is believed to exist only in Chinese deer farms, and both the plains bison and the chinchilla were spared almost certain extinction as a consequence of their commercial production. However, in today's computer terminology, the key words for game farming would be "rapid growth," "killing weights," "hybridization," "food conversion," and "hide quality," hardly synonymous with conservation. The preservation of the gene pool is of course important, but most of the animals now farmed have been so selectively bred and hybridized that they are already considerably genetically modified from their wild ancestors.

Game Farms

Deer are the most widely farmed game animals. Several have been domesticated or semidomesticated for many years, but in the last century there was an upsurge in interest in deer farming when it was realized that their gregarious grazing habits made them eminently suitable for farming pastorally like cattle and sheep. Until then, it has always been agreed that there were just two species of long-domesticated deer, the reindeer and fallow deer. However, at least one subspecies of sika deer (*Cervus nippon*), plus red deer, sambhar, and rusa deer, have been farmed in China for some time, possibly for several centuries. Then there are the semidomesticated red deer (*C. elaphus*), confined and controlled in English parks for many generations; and Pere David's deer (*Elaphurus davidianus*), extinct in the wild for many years and surviving only in a walled deer park in China, until shipped to Europe at the end of the nineteenth century. These animals have been joined by several other species, and deer are now the most plentiful of the newly domesticated herbivores. Although wapiti were farmed in the United States before the end of the nineteenth century, and musk deer in China in the 1950s or perhaps earlier, the modern era of large-scale deer farming began in 1970 in New Zealand. Now elk (*Cervus elaphus*), red deer, sika deer, musk deer (*Moschus moschiferus*), rusa deer (*Cervus timorensis*), axis deer (*Cervus axis*), and sambhar (*Cervus unicolor*), are raised on farms around the world, and their domestication is well underway. Their control and breeding, involving hybridization and selection for specific traits, such as increased antler size or body weight, faster growth, or better food conversion, are the very factors that result in rapid domestication.

In China, deer have traditionally been kept in small high-walled yards on brick or cobblestone floors, almost comparable to factory farming in the West, a form of wild animal husbandry that would not be well accepted outside the orient. Early in the last century travelers wrote of Kansu deer (*Cervus elaphus kansuensis*), one of the many Asiatic subspecies of the red deer, actually tethered in stalls, their hooves in a deplorable condition. The sambhar is also farmed indoors in Taiwan. Fortunately the western form of deer farming is pastoral, where the deer are kept like domestic cattle in grassy paddocks. New Zealand established the first pastorally based deer farms in 1970, and the system has since been followed in North America and Europe.

The major farmed species is the red deer, favored in New Zealand, Russia, China, Australia, Austria, and Germany. Elk are farmed mainly in North America, Russia, Korea, and China, and elk and red deer hybrids are raised in New Zealand. These hybrids reach killing weight faster than pure red deer, and with 30 percent heavier antlers than those of pure red deer at maturity. Sika deer are farmed in China, Korea, Japan, and Russia, as well as in New Zealand, where they have also been hybridized with red deer. Musk deer have been farmed for many years in China and Russia, and are now being kept in Korea. In Australia, red deer and elk are farmed in the temperate southeast, the chital or axis deer and the rusa deer are farmed in the tropical north. Excluding the elk, they are all established in the wild in Australia. Rusa deer are also being farmed in Mauritius and in Papua New Guinea. The long domesticated fallow deer is widely farmed, and to a lesser extent the reindeer also. Mule deer and white-tailed deer are farmed in North America, mainly for sport hunting. The major deer farming nations are New Zealand, with herds totaling almost two million animals, followed by China and Russia each with approximately 500,000 deer, and the United States with about 250,000, mostly elk. Australia has about 200,000 farmed deer, and Great Britain and Germany about 120,000 each.

The goals of deer farmers are the production of meat, antlers, and other by-products, such as testes, embryos, and penises, that have great value in oriental pharmacopeia. Since ancient times, deer antlers in velvet have been an important

Red Deer   *After centuries of partial control in deer parks, and more complete control on deer farms in recent years, the red deer is now a thoroughly domesticated animal. New Zealand is the major deer farming nation, with almost two million animals farmed for their antlers and their meat. The stubs of their removed antlers can be seen on the bucks above, behind a high fence on a New Zealand deer farm.*
Photo: Clive Roots

ingredient in traditional holistic oriental medicine, and are increasingly used in the West as a health supplement. Hardened antlers, finely ground and steeped in wine, or burned and the smoke inhaled, were believed to cure epilepsy; and, if sliced thinly while in velvet and steeped in herbs, are thought to offset the effects of snake venom. But by far the greatest demand for deer antlers, especially those of the sika deer, are in response to their reputation as powerful aphrodisiacs, and as rejuvenators believed capable of improving health and longevity. All possible measures to improve the quality of the deer and their products are therefore taken. To improve their stock, New Zealand deer farmers have imported bucks with "trophy antlers" from Europe, and stud farms have been developed, with show and sales rings reminiscent of those for thoroughbred horses.

To produce the perfect commercial deer, Pere David's deer were imported into New Zealand and crossed with red deer. Both male and female first-generation hybrids were fertile, and later studies on the carcasses of these hybrids showed less fat and more muscle, especially in the preferred hind quarters. Unfortunately, the imported deer also carried malignant catarrhal fever, and many of the original imports in the 1970s and 1980s died soon after arrival; this provided ammunition for the critics of game farming, of which there are many. In fact, it is a highly controversial industry that is opposed for several reasons, mainly its potential for disease importation and transmission to domestic animals, and in some regions, especially

North America, to native wildlife. For this reason, some state and provincial govern-ments prohibit the farming of red deer and elk, and with the federal governments, they are now looking more closely at the game farming industry regarding the future issuance of licenses. In New Zealand, escapee red deer and elk are of less conse-quence as they are both well established in the wild from past introductions, although the elk genes have been thoroughly overwhelmed by the more plentiful red deer. The exception is the sika deer, which cannot be farmed on South Island as it was never introduced there, whereas it lives ferally on North Island.

Paratuberculosis has already occurred in elk on farms in Alberta and chronic wasting disease in farmed deer in Saskatchewan, which was then transmitted to Korea when infected animals were shipped there in 1995. Also, in North America there is the risk of escapee elk, already likely to be subspecifically hybridized, cross-ing with local elk and diluting their genes. However, in total contrast, several states and provinces now recognize elk farming as a valid alternative livestock enterprise, with domestically bred animals being distinct from wild elk, their origin determined through DNA testing. Blood testing for purity is a requirement in states, such as Colorado, Montana, and Idaho, for another reason. They are concerned about escapee farmed animals mating with their native elk, so hybrids are not allowed entry.

The deer industry in New Zealand is now very large and very sophisticated. About 1.9 million deer live on 4,500 farms, with 90 percent of the animals either red deer or red deer/elk hybrids, and their production and marketing is controlled by Deer Industry New Zealand, established in the mid-1980s. Since 1993 venison has been marketed under the "Cervena" appellation, which can only be used by franchised New Zealand exporters who meet strict quality and marketing standards. Venison is favored by many people for its flavor, texture, and low cholesterol con-tent. Germany is the main buyer of New Zealand venison, while Korea purchases 90 percent of the antler crop. Antlers are cut in the growing stage while they are in velvet, and are exported in "whole stick" form to Korea, whose regulations prohibit the entry of sliced velvet. Russian and Chinese deer farmers also supply many antlers for the oriental trade, where the larger elk or hybrid deer antlers are favored, due to the perception that "bigger is better." North America is a relatively new supplier of antlers in velvet to Korea.

Unlike the many species of deer that have proven conducive to farming and domestication, only one antelope—the eland—has been completely controlled and domesticated, because it is the only species for which a serious farming effort has ever been made outside a species' natural range. It is therefore the classic exam-ple of antelope farming and domestication, and it occurred at Askaniya Nova on the grasslands of the Ukraine, where almost 1,000 calves have been born since the first two pairs of eland were obtained from East Africa in 1892. Farming eland in its natural habitat is considered of greater potential value, to provide protein for the natives; but attempts to domesticate the eland there, on the Galana Ranch, were unsuccessful, as they preferred to feed at night when it was cooler, but when they were vulnerable to prowling lions and hyenas.

The Galana Ranch in southeast Kenya has had more success in farming the fringe-eared oryx (*Oryx gazella callotis*). Its 1,930 square miles (5,000 sq km) of semiarid bushland has low rainfall and high temperatures and can support cattle only for a few months annually after the rains. A semidomestication program was started three decades ago for the oryx, which were considered more suitable than eland because they feed during the heat of the day and can then be penned at night to protect them from lions and hyenas. The oryx undergoing domestication are caught in the wild as calves and penned in herds to accustom them to artificial foods and the sights and sounds of human activity. They have bred readily, their health is good, they are generally resistant to disease—especially trypanosomiasis, the scourge of domestic cattle in the region—and, as they rarely carry ticks, they are not susceptible to tick-borne diseases. They have evolved to make full use of the coarse local herbage, digesting the native grasses more efficiently than cattle, but when they are penned they are fed local cut grass and lucerne. They need water only every two to three days during the hottest time of year, whereas cattle need to drink daily. When the grass is green, they may not drink for several weeks. They seem destined to be one of the most successful of the new domesticates on marginal lands. This form of wildlife control is a combination of ranching—where little or no control or feeding is provided—and farming, where the animals are completely controlled, and most of their food is provided.

The wild cattle and their relatives the bison and buffalo have not been subjected to human control with the same enthusiasm as deer, and only one species, the plains bison, can really be grouped with the long-controlled cattle, water buffalo, and yak as a newly domesticated species. The plains bison is the most farmed game animal in North America, with an estimated half a million animals on farms and ranches. Although zoological gardens have been keeping musk oxen, gaur, banteng, European bison, and African buffalo for many years, only the musk ox is being farmed, and only in a minor way. Despite their evolution in a very harsh environment, musk ox have thrived in some regions of the United States and have done particularly well in the San Francisco Zoo. In view of the great value of their wool, known as qiviut, it is surprising that more attempts have not been made to farm them. They are simple herbivores, needing a diet of coarse herbage, but in confined areas they are very rough on fences, and mature bulls have even challenged concrete walls to butting contests. There are several farms in Alaska, where they are tourist attractions as well as qiviut producers. The leader in commercial musk ox farming is the Musk Ox Development Corporation, which has a farm at Palmer, Alaska. Their animals are brought into barns in the late spring to be combed, and the qiviut is hand-knitted into garments by native Alaskan women of the Oomingmak Musk Ox Producers Cooperative. The musk ox's downy undercoat is very durable and does not shrink. It is several times warmer than most wools, and is so clean it does not need carding before it is spun. An adult animal yields about 5 pounds (2.2 kg) of qiviut annually, but only a few ounces are required to make a scarf.

Unlike the other game-farmed animals, which are all herbivores, the European wild boar is an omnivore, that will eat just about anything of plant or animal origin. The ancestor of the domesticated white pig many millennia ago, it has recently undergone redomestication again but this time as a pure animal, characteristic of its Eurasian ancestors. It is now farmed in many countries for its lean meat.

Fur Farms

Farming wild animals for their furry skins is believed to have begun in the United States when mink were farmed soon after the Civil War. Blue foxes were farmed in Alaska in 1885, and silver foxes on Canada's Prince Edward Island in 1892. These early beginnings gave rise to many mink and fox farms in Europe and North America in the early years of the last century. Norway began farming silver foxes in 1914, and Finland and Denmark developed mink farms a few years later. In those days foxes were generally kept in small enclosures, and one of the most innovative forms of farming occurred in 1917, when blue foxes were allowed to run free on several small islands in Alaska's Prince William Sound. They were fed daily, and then trapped at their feeding stations in December when they were 10 months old and had their prime winter coats. Chinchilla farming began in the United States in 1923, while nutria or coypu were first farmed in Argentina in 1922, and then in England in 1929. Since those early days, fur farming has become much more intensive, with the animals being mostly kept in small wire cages in conditions that have been likened to the factory farming of domestic livestock, and this has been one of the major criticisms of the industry. The animals currently involved in the international fur trade are mainly mink and foxes, plus chinchillas, coypu, raccoon dogs, sables, and ferrets, and to a much lesser extent lynx and bobcat.

Before the era of fur farming, the world's garment industry was totally dependent for its furs on trappers who set steel-jawed traps for mink and beaver in Canada's north, for sable and raccoon dogs in Siberia, and who virtually exterminated the wild chinchilla in the Andes. The position is now reversed, with farmed animals providing most of the world's furs. In the century that has passed since farming began, growth has been steady despite periodic swings. The anti-fur campaigns of the 1980s and 1990s severely affected the fur farm industry, but it has rebounded, and furs are now more popular than ever. Fur wearing, and consequently fur production, has made a comeback, and about 88 percent of all the furs used annually in the production of garments now originate on fur farms. Fur farming has therefore considerably reduced the trapping of wild animals, but there is a continual battle of words between the fur industry and activists. Actions also, in the form of the liberation of animals from their cages.

Pressure from anti-fur farming activists and organizations has resulted in the banning of fur farming in several countries recently. It was banned in England and Wales in 2000, and Scotland followed suit in 2002 to prevent farmers relocating there, although there were no farms in Scotland at the time. Fox and chinchilla farming was banned recently in Holland and Sweden, although mink farming continues in Holland. Mink farming is banned in most of Austria, and the requirement to provide ponds for their caged animals has resulted in the remaining farmers there concluding that mink farming is no longer viable. Sweden and Germany plan to introduce similar "pond legislation," which requires farmers to provide mink with a pond 78 inches square by 20 inches deep (2 m square x 50 cm), which would actually be larger than their approved cage size. Fur farming is already banned in the German states of Hessen, Bavaria, and Schleswig-Holstein. Despite the bans, however, fur production continues to increase. The anti-fur-farming brigade's claims that the trade is diminishing is not supported by world production figures, nor by the USDA's reports for 2004 and 2005. In the United States, mink production increased in 2004 by 1 percent over 2003, the crop being 2,563,000 pelts, even though the number of farms declined from 305 in 2003 to 296 in 2004. In 2005 production increased again, up 3 percent to 2,628,000, yet the number of farms dropped to 277. In Canada's province of Newfoundland and Labrador, the number of breeding mink increased from 1,000 in 2001 to 22,000 in 2005, with a potential production of 100,000 pelts annually. The world production of mink in 2005 was 40 million skins, and of foxes 6.5 million, both major increases over previous years.

Europe is the center of fur farming, containing almost 75 percent of the world's mink farms while North America has 12 percent, the balance being mainly in Russia and China. In 2004 there were 6,500 fur farms in Europe, and 1,135 fur farms in North America. However, China is the growing powerhouse in fur production, not only of "wild" species but also of cats, dogs, and rabbits, and there has been a tremendous increase in the production of furs and fur garments in recent years. However, Chinese furs are generally considered of lower quality, and a large percentage is "budget" fur, used for trimming hoods and collars. Most of the world's mink farms are in Denmark, which produced 12.9 million skins in 2005. In the same year China produced 8 million skins, the Netherlands 3.3 million, the United States 2.7 million, and Russia, Finland, Canada, and Poland about 2 million skins each. China is the world's leading producer of fox skins, with 3.5 million marketed in 2005, followed by Finland with 2.2 million. Norway, Russia, and Poland are also major producers of fox pelts. In 2005 China also produced the most raccoon dog pelts, 2 million of them, again followed by Finland, with just 110,000. The United States is the world's major producer of chinchillas, followed by Argentina. Russia produces 150,000 sable skins annually, but apparently trappers there are now harvesting even more from the wild. Russia has guarded its sables for generations, and live animals could not even be exported to zoos. Nutria are currently farmed in Poland, the Czech Republic, and Armenia. Ferret farming has declined in recent years, with Finland now leading world production with 11,000 pelts per year.

The international fur trade is a massive industry. One million people are employed worldwide, from farm to salon. The European Union in 2004 employed 106,000 people full-time and 108,000 people part-time in the fur industry. The Russian fur trade is said to be currently worth $2.5 billion annually, and in the United States the trade at the retail level averages $1.5 billion in annual sales. The 12.9 million mink skins produced by Denmark in 2005 ranked third in export value after bacon and cheese, and they are a major item of the export trade of both Finland and the Netherlands. Farms vary in size, and in the United States most fur producers are smaller "family farms." But there are a few very large ones, such as the Wisconsin farm that produces 100,000 pelts annually, and that state leads the nation in mink production. Most carnivorous fur bearers receive a paste made from the by-products of human food production. In 2002 fur farmers in the European Union fed their animals 640,000 tons (650,000 tonnes) of poultry, fish, and slaughter-house by-products or offal—fish heads, viscera, intestines, literally anything unsuitable for human consumption. In North America 200,000 tons (203,000 tonnes) of by-products are fed by fur farms. The mink on the large Wisconsin fur farm receive

2 million pounds (907,000 kg) of expired cheese, and 1 million pounds (457,000 kg) of damaged eggs annually. The use of such foods is considered a side-benefit of fur farming, keeping the cost of human food production down. Consequently, despite the pressure from activists, some governments are reluctant to interfere with the industry and the revenue it generates. However, neither the jobs, nor the revenue, nor the use of offal can compensate for cruelty and suffering.

In the modern era there has always been opposition to the wearing of fur. In the heyday of wild trapping, the concern was for the suffering caused by the capture methods. Today it is mainly concern for the intensive nature of the animals' housing, and the methods of killing, that are considered cruel. Many people consider fur farming inconsistent with the proper values and respect for wild animals, as fur bearers are still considered by some to be wild and therefore unsuitable for farming. In 2001 the European Union's Scientific Committee on Animal Health and Animal Welfare (SCAHAW), concluded that practically all fur-bearing animals—mink, foxes, ferrets, raccoon dogs, chinchilla, and coypus—were unsuitable for life in captivity. It was surprising to see ferrets included in their list, considering they have been domesticated for about 2,000 years. Foxes, too, in view of the extensive selection experiments conducted in Russia between 1940 and 1980, that proved conclusively that silver foxes can not only be domesticated very quickly but are responsive to selective breeding for tameness, and after just a few generations were actually seeking human company. What is rarely appreciated, however, is that domestication is not necessarily synonymous with tameness, but with acceptance of man's control. But where fur-farmed animals are concerned, domestication continues to be an issue.

Surprisingly, of all the animals that are now controlled by man—by the breeders of pet trade animals, by turtle farmers, crocodile farmers, deer farmers and even

Arctic Fox   Fox farming began at the end of the nineteenth century in Canada, and they are now second only to the mink in annual production. China is the main fur-farming nation, producing 3.5 million pelts annually, followed by Finland, with 2.2 million. Both the red fox and the Arctic fox are farmed, and hybrids between the two species have been produced through artificial insemination.
Photo: Clive Roots

zoo keepers—the question of an animal's domestication has become an issue only where fur farming is concerned, and only regarding the animal's suitability for intensive care. The ongoing argument over whether fur-farmed animals are domesticated or not is purely for the purpose of condoning or discrediting their means of housing. Specifically, that if they are not domesticated and conditioned to control, then they will be distressed by close confinement, and their intensive care is therefore cruel. Conversely, it is acceptable to keep domesticated animals in cages that are shorter than the animal's body length. Naturally, opinions on the domestication of fur-farmed species offer two completely opposing views. Critics of the fur industry maintain that mink and foxes are not domesticated animals, and do not condone their keeping anyway, just as they oppose the factory farming of calves and chickens. In contrast, the industry maintains their animals are domesticated, and consequently their housing is adequate. Whatever the status of fur-farmed animals, the long-term confinement of any animal in such tiny cages is unacceptable.

The International Fur Trade Federation, which represents thousands of fur farmers in 23 countries, believes that after 100 generations, fur farmed animals are effectively domesticated. They consider the farmed mink's temperament to have changed from that of a nervous, agitated animal to one that now reacts with curiosity and inquisitiveness to the approach of humans. The fur industry considers their animals to be healthy, breeding well, and generally docile. However, Oxford University biologists who studied the behavior of captive mink believe they are not domesticated, despite the many generations raised on the farm, and that they suffer from confinement, especially if they are unable to swim. The European Union's SCAHAW, in a report published in December 2001, also concluded that the animals currently being farmed cannot be considered long-domesticated, and largely "retain the characteristics of the wild animals."

Freshly caught mink, like most wild animals when they are first caged, were undoubtedly stressed in the early days of fur farming; but now, after so many generations, it seems unlikely. Researchers at the University of Copenhagen's Institute of Population Biology developed blood tests to identify the concentration of white blood cells in the mink's blood as a measure of long-term stress, and concluded that the animals on Danish farms were not suffering. The SCAHAW Report, however, concluded that mink and fox, unlike other farm animals, have experienced very little selection for characteristics other than fur production; and that there has been a limited amount of selection for tameness or adaptability to a captive environment. Mink have been selectively bred for almost a century, which may or may not mean 100 generations but must certainly be close to that figure, as future breeding stock are normally held back each year. They have been selected for fur color and quality, but obviously also for tameness and calmness, either intentionally or unintentionally. It is inconceivable that nervous and highly stressed animals would breed well or would even be retained for breeding. The industry says that fur farmers are in fact selecting for tamer, more confident animals. Stressed or not, the inhumane treatment of animals, of any species, can never be tolerated, and housing them in such tiny cages is indeed inhumane. Domesticated animals are completely dependent upon humans, who therefore have a moral and ethical duty to provide the best possible care.

Tamer animals are actually not all that difficult to produce. Russian scientist Dmitri Belyaev (1917–85) began breeding the silver fox (*V. vulpes*) in 1940 and for almost 40 years, during which time he bred 10,000 foxes selected only for friendliness, breeding only animals that showed the most positive response to humans. After just a few generations, his animals sought human contact; and after 18 generations of selective breeding, his foxes showed many of the characteristics of the domesticated dog, with curly tails, drooping ears, and shorter muzzles. They showed no fear of humans, and actually enjoyed human contact and petting.

Opposition to the wearing of fur and disagreement over their position in the domestication process aside, criticism of fur farms stems mainly from their animal housing and killing methods. Photographs show farmed mink in Europe and North America receiving very considerate care, but it is the housing and its effect on the animals, rather than the care they receive, that are of greater concern. For every photo showing a seemingly contented mink raising her kits in her straw-lined nesting box, there are numerous photos of the horrific conditions on Chinese fur farms. There the animals are also kept in the typical small wire cages, but the mink are not provided with nest boxes, nor the foxes with sleeping platforms, and they are often outdoors without shelter, completely exposed to the elements. It is obviously not in any farmer's best interests to lose animals through poor husbandry, and it is certainly not for the want of regulations. In the European Union, Council Directives lay down rules concerning the welfare of all farmed animals and their killing. In the United States, all farms are regulated by the USDA and must also meet state requirements to ensure high-quality animal husbandry. Russian fur farmers are covered by specific laws and agricultural regulations. In China, however, there is a total absence of animal welfare regulations.

Space is important for confined animals, but as zookeepers are aware, even the largest cage is useless unless it provides for the animal's behavioral and physical requirements. It must permit walking, running, sleeping, hiding, climbing, swimming, and even burrowing, as appropriate. The SCAHAW Report was critical of the fur industry's current animal husbandry methods and identified major shortcomings in cage size, management methods, and the lack of staff training and handling methods, and concluded that the cages do not provide for the animal's needs. Where mink were concerned, they said it prevents normal animal behavior—swimming, running, and climbing. Austrian farmers concluded that providing such "swimming water" for their mink would be uneconomical.

Farmed mink and foxes are obviously not kept in conditions that are compatible with their natural requirements. Nor can they be, if they are expected to be commercially viable. The recommended cage size for a single mink is 24 inches (61 cm) long by 15 inches (38 cm) wide by 14 inches (36 cm) high. Cages are made of wire mesh, and have a mesh floor that is preferable for hygiene purposes as the feces drop through. A mother with kits may have a cage 6 inches (15 cm) longer, with a small, attached, straw-filled nest box. Foxes and raccoon dogs have similar wire mesh cages providing slightly more than double the mink's space, and they are usually provided with a shelf above the mesh floor on which they can sleep, and a nest box for whelping. These cages are normally in an open-sided building, its roof protecting the animals from rain and sun. Mink and foxes are very hardy creatures, and are more likely to suffer from heat stress than from cold. An adult male mink has a head and body length of about 21 inches (53 cm), and its tail is 7 inches (18 cm) long, so from nose to tail tip it is actually longer than the recommended cage, in which it has to spend its whole life, although that is relatively short, except for those kept as future breeders, which get the 6-inch-longer (15 cm) cage.

Abnormal behavior is usually an indication that an animal is stressed or is facing other problems concerned with its welfare. The lack of space, lack of seclusion, exposure to noise, unavoidable human activity, and continual contact with neighboring conspecifics for normally solitary species, which both foxes and mink naturally are, all have the potential for stress. So does an animal's inability to hunt and to establish and protect a territory. Critics of fur farming claim the animals are in a constant state of stress and fear, and that they are psychologically damaged. They claim there is a 20 percent mortality of mink kits, that the animals exhibit stereotyped behavior consistent with severe stress, and that they also commit self-mutilation, especially tail chewing. It is claimed that small cages result in excessive fear of humans, which has not been alleviated by genetic modification or long contact with humans. Foxes are said to suffer limb weakness from lack of exercise, and there is a high rate of breeding failure—due to low conception rates and poor survival of the young. The animals are said to be diseased and heavily parasitized, and that contagious diseases are likely to pass from animal to animal. While the latter is certainly true, as it is for any collection of animals, the fur farms have a regimen of vaccination and de-worming, and it is unlikely that any Western farmer would jeopardize his livelihood with such irresponsible husbandry. Fur farmers and their veterinarians insist that the animals are healthy, happy, and well adjusted to their environment.

The industry considers its fur-bearers to be among the world's best cared-for livestock, and believes their cages now provide sufficient space for normal movement and investigative behavior, even though they are shorter than the adult mink's body length. Its recent environmental enrichment of the cages includes providing nest boxes for the mink and observation platforms for the foxes. It claims the animals have a good diet, designed to produce a lustrous pelt, and are provided with the best veterinary care, all to protect the farmer's investment. The fact that their animals breed regularly is cited as proof of their contentment, but breeding is by no means an indication of perfect conditions, as the size of its accommodation has very little effect upon an animal's reproductive powers. Unfortunately, the trend nowadays is to provide the absolute minimum space for animals, whether they are macaws, pheasants, marmosets, or leopard geckos; as many have shown, they will breed in the smallest of cages, and the fur industry's recommended cage sizes are woefully inadequate.

The other major criticism of fur farming concerns the killing methods. Fur-bearers are killed humanely in North America and Europe, mainly by gassing or electrocution. Carbon monoxide is released into a "gas box" containing the animals, resulting in their death in about 30 seconds. Electrocution involves inserting an earth probe into the animal's anus, and then a charging probe into its mouth, and it is electrocuted when it bites the probe. Both methods have been attacked by activists as being frequently conducted by untrained staff resulting in suffering, plus claims that the gas boxes are merely attached to a vehicle's exhaust, which does not kill the animals quickly. In China, it is reported that foxes are held up by the legs and then beaten on the head with an iron bar; and that many are then skinned while they are unconscious.

However, despite the very definite and opposing views on fur farming and the animals state of domestication, the attitude of many people to fur farming, like other forms of farming, is one of acceptance. Most do not wish to hear about an animal's lifestyle before it produced the products on the butcher's block or in the fur salon. They believe it makes little difference if the farmed animal is a pig, a chicken, or a mink, provided its welfare was supervised and that it was humanely killed. In the Netherlands, 71 percent of people surveyed in 2000 agreed with this philosophy, and in another survey 67 percent of Scottish people also supported it. In the United States 86 percent of people surveyed supported an individual's right to wear fur, and 69 percent of Finns have a positive attitude to fur farming. The domestication of mink and foxes, and more recently of other fur-bearers, will undoubtedly continue.

Other Farms

In regions of Africa where the use of bushmeat has traditionally been a major source of food, there have been efforts to farm some of the animals which have provided this protein. Market prices for these favored meats are usually higher than for the meat of domesticated cattle and goats, and the demand has led to overkill of many of the smaller mammals—the only ones surviving in many regions of West Africa—such as duiker, monkeys, and large rodents. While farming duiker or monkeys for meat is certainly not viable, the prospects of farming rodents offers a cheaply produced alternative to conventional livestock production and a reprieve for free-living wild animals. Two large rodents—the giant cane rat or grasscutter rat (*Thryonomys swinderianus*) and the giant pouched rat (*Cricetomys gambianus*)—were considered to have great potential. The value of producing native herbivorous animals is that their foods—mainly grasses and reeds—can be gathered locally, and bush farming must have low production costs and low output of effort to be successful. Farming native rodents that the people are familiar with, and have traditionally hunted and eaten for generations, is considered a valid means of replacing the protein supplies now lost through overhunting and loss of habitat. Rabbits would be easier and quicker to produce, but they are not familiar animals to the people of West Africa, whereas there is an unlimited market there for native rodent meat. Chickens would also be more productive, but it is pointless farming animals that need grain to produce protein.

The giant pouched rat has been traditionally trapped for food in rural areas of West Africa, and a program to domesticate it was begun by Nigeria's University of Ibadan in the early 1970s, in an effort to maximize meat production from the species. Wild-caught rats adapted to captive conditions very quickly, readily ate replacement diets, were breeding within two months of capture, and after four months had lost all the antagonistic behavior characteristic of the wild rats. The rat is still bred at the university and is considered domesticated, but its commercial production did not proceed due to superstition and cultural aversion by many tribal groups. The domestication of the giant cane rat or cutting grass rat was more successful, and it is now farmed in several West African nations.

Wild civets, members of the mongoose family *Viverridae,* have traditionally been trapped for their musk, but efforts are now underway to farm them. In China's Guangdong province, many farms keep the masked palm civet (*Paguma larvata*), an animal suspected of involvement in the epidemic of Severe Acute Respiratory Syndrome (SARS), the respiratory infection that caused the death of 774 people in 2002–3. In Ethiopia the African civet (*Civettictis civetta*) has traditionally been the main supplier of musk or civet, a product of its anal glands, to the world's perfume industry, but they were killed for their musk. Farms are now being established in Ethiopia to breed civets and extract the musk periodically, sometimes every two weeks. One farm claims to have 400 civets that produce 1,600 pounds (725 kg) of musk annually, a seemingly excessive amount, which is exported for $250 per pound. The government has apparently developed a code of practice, with standards of care and testing methods, as civet from trappers in the past has been extended with vaseline. Even so, the civets are apparently kept in deplorable conditions, and the frequent musk extraction is a traumatic experience for them. Fortunately, some perfume manufacturers now use only synthetic musk.

■ SOME OF THE SPECIES
Red Deer (*Cervus elaphus*)

Red deer are widely distributed in Europe and central Asia and are the only true deer occurring naturally in Africa—in Tunisia and northeastern Algeria. They have been successfully introduced into Argentina, Chile, New Zealand, Australia, and the United States. They are adaptable animals and thrive in a wide range of habitat, including grasslands and temperate forests, on alpine slopes and in very arid regions. Red deer belong to the same species as the elk (*Cervus elaphus*) of North America, but their subspecies are generally smaller. Their wide natural range across Eurasia has resulted in the evolution of almost 20 distinct races, which differ considerably in body and antler size. On Corsica and Sardinia, stags of the subspecies *C. e. corsicanus* weigh only about 250 pounds (113 kg) and have small antlers. In Scotland *C. e. scoticus* stags weigh about 300 pounds (135 kg), and their antlers 13 pounds (6 kg); and the North African race *C. e. barbarus* is similar. The largest red deer occur in Poland and Romania, where stags of the race *C. e. hippelaphus* can weigh 640 pounds (290 kg), rivalling elk in size, and have antlers weighing 22 pounds (10 kg).

Red deer can hybridize with elk and produce fertile offspring, and they can also cross breed with axis deer, Pere David's deer, sika deer, and hog deer, with varying degrees of fertility. Throughout their natural range and in their new countries, the red deer's genes have consequently been contaminated by those of several other species. In Great Britain and Texas they have hybridized with the introduced sika deer. The sika deer, imported into Scotland as park animals towards the end of the last century, escaped or were released and have mated with the native red deer, and the continued spread of hybridism now affects a large part of the country's red deer population, with 40 percent of the animals in Argyll having sika deer genes. The "mongrels" are smaller-bodied animals with smaller antlers, and are considered a threat to tourism and to the shooting trade.

Red deer were introduced into New Zealand from English country parks in the years between 1851 and 1923, and became established on all three major islands. Then in 1905 wapiti were released in Fiordland and hybridized with the red deer. These animals flourished in New Zealand's mild climate, lush forests, and in the absence of predators. During the first 50 years in their new land, they gained in body and antler size, but this could not be maintained when the best feed was utilized and the population became very large, causing great damage to the native forests. This led to an eradication campaign and the marketing of the meat and antlers, and their reduced numbers then prompted farming in 1970 to continue supplying the developing market. The first farms were stocked with wild-caught deer, then imports were made from Europe to improve the quality of the stock. Red deer are now the most widely farmed species in New Zealand, where they form about 90 percent of the country's 1.9 million captive deer. They are raised as pure animals and are hybridized with imported elk to produce larger offspring with much thicker antlers. Castrated stags have grown even larger antlers, double the thickness of normal, but are unsuitable for the oriental velvet antler trade as they are considered "eunuchs" and are therefore unacceptable for virility potions. Most of the venison is exported to Europe, while the antlers are shipped to South Korea and Hong Kong.

Elk (*Cervus elaphus*)

Elk or wapiti are the North American equivalent of the red deer, and belong to the same species. Excluding the Tule elk, their subspecies are larger than the Eurasian subspecies of the red deer. Their natural, recent distribution was as follows:

Eastern elk (*C. e. canadensis*), formerly of eastern North America, but now extinct, and some of its habitat has been repopulated by *C.e. nelsoni*.

Roosevelt elk (*C. e. roosevelti*), of western North America (Oregon to southern British Columbia and Vancouver Island). This is the largest subspecies, in which mature males may weigh over 1,000 pounds (543 kg).

Rocky Mountain elk (*C. e. nelsoni*), of Alberta and British Columbia south into Wyoming, and introduced into northern Alberta, southern Yukon, and Ontario.

Manitoba elk (*C. e. manitobensis*), a native of Manitoba and Saskatchewan, which only survives as a wild animal in reserves and national parks.

Tule elk (*C. e. nannodes*), the smallest and rarest species, occurring only in California.

Elk are large deer, the second-largest species after the moose, and they have heavy bodies, thick necks, and slender legs, with a tawny-colored rump patch.

Their enormous many-tined antlers may weigh 40 pounds (18 kg), and the main beam may measure 59 inches (150 cm). After a gestation period of 246 days, the elk hind gives birth to a single fawn, or occasionally twins. The fawn is quite pre-cocial and soon stands to suckle, then remains hidden in tall grass or bushes, its mother keeping watch from a distance and returning to feed it periodically. They are recent additions to game farming, being first farmed in North America in the late 1960s, but have already become the favored species. There are currently approximately 140,000 animals on farms, but are considered too valuable to kill for their meat, and for now are used as breeding stock and for nonterminal antler production. Meat production is considered unlikely to happen until the present population doubles in size. Their antlers are removed about 70 days into their growth period, when they are "in velvet"—when they have a soft covering and are richly supplied with blood vessels. Most antlers are shipped to South Korea, whose people believe they have valuable properties for rejuvenation and long life. However, products containing antlers are increasingly being marketed in the United States. Grade A antler is currently worth $35 per pound, and depending on their age and condition stags can produce between 20 and 40 pounds (9 to 18 kg), at each annual cutting. In New Zealand, elk were only released once, in Fiordland in 1905, but their genes have since all been diluted by cross-breeding with the more plentiful red deer that were introduced earlier, but toward the end of the century many were imported for farming.

Musk Deer

Musk deer have been farmed in China for at least 50 years and probably a lot longer. Three native species are involved, but most farmed animals belong to the species known as the forest musk deer (*Moschus berezovskii*). It is a small deer, when mature weighing up to 24 pounds (11 kg) and standing about 21 inches (55 cm) high at the shoulder. It is an antlerless species, the males having instead elongated upper canine teeth that project almost 3 inches (7.5 cm) below their top lips, which are used in settling territorial disputes with other males. When mature, the musk deer stag has a musk gland or pod which contains almost 1 ounce (28 g) of musk, a coveted ingredient in the perfume industry. A very expensive commodity, it currently has a retail value three times that of gold and is in great demand. France, Germany, and Switzerland are the major importers of musk for the perfume indus-try. Russia supplies much of the world's musk from its native wild species the Siberian musk deer (*M. moschiferus*), but there is still a large illegal trade and the wild populations are being rapidly reduced, despite their listing on Appendix I of CITES, which prohibits commercial trade in vulnerable species or their products.

The amount of musk entering international trade annually represents the glands of almost 10,000 mature males, and as shooting and snaring are the main means of capture, and females and immature "pod-less" males are also killed, the annual kill is believed to be at least double that number, which represents a large percentage of the population. Chinese musk exported to Hong Kong is usually in gland or pod form, which means the animals were killed to obtain it. China's wild musk deer population has therefore also been seriously depleted, and is down to about 200,000 animals from its original three million in the 1950s. Unfortunately, the farming of musk deer has done little to reduce the large annual kill, mainly because it has not expanded like the farming of sika deer. There are currently believed to be only about 2,000 musk deer on farms, where they are kept in walled yards, in the traditional Chinese manner. They are far more secretive about their musk deer farms than other deer farms, however, and I was refused entry to one of their largest farms in Sichuan province. These captive animals produce only a fraction of the current world demand, which is estimated at 2,204 pounds (1,000 kg)

Eland (*Taurotragus oryx*)

The largest of all antelope, the eland is an ox-like animal with a tan coat (females) or a dark tan with a bluish tinge (males), and young animals have cream-colored vertical stripes on the upper body. It has a hump on the withers and a dewlap framed with long black hair, and both sexes have heavy spiral horns. The eland's walk is accompanied by foot-clacking, believed to be caused by tendons snapping back when the foot is lifted, or by their large hooves clicking against each other after spreading to take the animal's weight. Males may stand 5 feet 10 inches (1.78 m) at the shoulder and weigh up to 2,000 pounds (900 kg), while the females reach a height of 4 feet 8 inches (1.42 m) and weigh up to 1,000 pounds (450 kg). The common eland is a very social animal, living in herds on the plains and wooded savannah of eastern Africa, where it is a grazer of coarse herbage and browses the leaves of trees and shrubs.

The eland is the only farmed and now domesticated antelope. The herd at Askaniya Nova in the Ukraine was established in 1892, with four animals imported from East Africa, and is considered a major intentional domestication project for a new species. The eland are kept just like cattle, and are milked daily. Records of their milk production have been kept since 1950, and selection has been made on the basis of yield and fat and protein content. Eland milk has double the solids of regular cow's milk, and is richer in calcium and phosphorus. With good nutrition, the cows at Askaniya Nova lactated for up to 300 days, some animals providing up to 1.8 gallons (8.5 L) of milk daily. Although most African antelope have subcutaneous fat, eland meat has marbling through the flesh, like beef, and remains relatively tender into old age.

The Askaniya Nova project has proved that eland can be easily tamed and herded, and even the bulls are usually quite tractable. However, this was not without some cost, as considerable inbreeding occurred due to the small number of founder animals. The offspring produced prior to 1937, when an unrelated male was obtained, were sired by the original two males and their male offspring. By 1930 serious signs of inbreeding were obvious, with calves suffering prenatal mortality, weakness at birth, and serious skeletal and hoof deformities. New blood was also acquired in 1959 and 1964 to outcross the herd, which is now being selectively bred for meat and greater milk production. Attempts have been made to cross eland with domestic cattle, both at Askaniya Nova and also previously in England in the mid-nineteenth century; but they were unsuccessful, as to be expected between such distantly related species. However, domestication does not necessarily involve

Eland *The largest antelope, and the only truly domesticated species, eland have been farmed at Askaniya Nova on the Ukraine steppe since 1892. They are kept just like dairy cattle, milked daily and their production recorded, confined in winter to avoid the subzero climate, and fed completely artificial diets.*
Photo: Susan Adams, Shutterstock.com

acclimation, and the Ukrainian eland cannot withstand the rigorous winters of the steppe. They are brought into their indoor pens at night when the temperature reaches 55°F (12.7°C) and are housed permanently indoors for the winter when the temperature drops to 38°F (3.3°C). They are let out again in spring.

Plains Bison (*Bison bison*)

Bison are the most familiar of the wild cattle; huge dark-brown animals with a chest mane and chaps of long hair on their front legs down to the pastern joint. Their large chests and massive skulls are responsible for their heavy forequarters, and the extension of the last cervical vertebra and the first 10 thoracic vertebrae give the bison its characteristic hump. Their small pelvis accentuates the heavy front-loading effect. As in all land mammals, color mutants occur and gray, piebald, and albino bison are occasionally seen. Adult bulls stand 6 feet (1.83 m) at the shoulder hump and may weigh 2,000 pounds (900 kg). Cows are about 1 foot (30 cm) shorter and reach a weight of about 1,200 pounds (550 kg). Their gestation period is nine months, and the calves weigh about 48 pounds (22 kg) at birth and are quite precocial, running with their mother within hours of birth and beginning to nibble grass soon afterward. Bison are 15 percent more efficient at converting forage to meat than cattle.

Bison are commonly known as buffalo, a misleading name as wild buffalo live in Africa and the wild water buffalo still occurs in southern Asia, with domesticated descendants in many countries. From their original distribution on the North American plains and lower mountain slopes, the demise of herds estimated at 100 million to a few hundred has been frequently repeated. They have now recovered, with a population of over half a million throughout North America, although only the race known as the wood bison lives in a manner similar to the free-ranging state of old, subjected to the pressures of natural predators, but unfortunately heavily infected with tuberculosis and brucellosis. With the exception of the wood bison, the captive population of bison can be considered virtually domesticated, even though this process only began early in the last century when the survivors were gathered together for captive breeding. They have since been totally controlled by man, with culling an integral aspect of their management, even in the largest herds in protected areas.

The nomadic plains Indians were dependant upon the bison. They ate every edible part and cached supplies for winter, either by freezing it outdoors, or by dehydrating it into jerky, mixing it with buffalo fat, berries, or maple syrup, and sealing it in rawhide bags. Modern man has also recognized the value of its meat, and the bison is currently the major species in the commercial exotic animal industry in North America. Bison meat is very lean and dark red due to its lack of marbling and high iron content, and has more protein than beef. The low fat content of ground bison produces a rather dry hamburger, which for some is more than compensated for by its low cholesterol. Whereas USDA choice lean beef has 72 mg of cholesterol per 100 g, bison meat has 62 mg per 100 g, similar to chicken breast. Bison meat is very tender, especially the typical choice beef cuts such as tenderloin and rib-eye; but of course tenderness is dependant upon age, and for the meat trade bison are normally slaughtered when between three and four years old, when a bull can weigh 1,300 pounds (600 kg). Another important factor to health-conscious people is the fact that the bison's diet generally does not include additives such as hormones, growth stimulants, and antibiotics. However, bison are synonymous with the open range. They are exclusively grazers, thriving on low-quality forage, so their evolutionary physiology makes them poor subjects for the intensive conditions of the feedlot. Due to their size and great strength, very sturdy fencing is required, especially in confined areas such as yards and shelters. Their docility is deceptive, for they are unreliable and very agile for their size, charging quickly and keeping pace with a horse.

Bison account for the major part of the game industry in North America, far more than deer farming. There are said to be about 280,000 in Canada, and almost as many in the United States, where the current bison meat production is 7.5 million pounds (3.4 million kg) annually. Many attempts have been made to hybridize bison and cattle, to produce animals known as "cattalo" or "beefalo," but this was largely abandoned about three decades ago due to infertility problems. Like mules, hybrid male calves were generally infertile. A few are still being bred, and have 3/8 bison blood and 5/8 cattle blood. Bison have also been crossed with yak, the calves being known as yakalo.

American Mink (*Mustela vison*)

The mink is a large weasel, with males weighing up to 34 ounces (1 kg), while the females are slightly smaller. It has a natural range in North America from Alaska and Labrador to the Gulf of Mexico, but is now established as a result of escapes from fur farms and deliberate introductions in Argentina, Iceland, Great Britain, and throughout Europe and Soviet Asia. In 1933 American mink were released into the wild in Russia to begin a new fur industry; and in recent years the wild (feral) population has increased as a result of activists releasing farmed animals. The native species in Eurasia, the European mink (*Mustela lutreola*) is now very rare due primarily to competition from the alien mink.

The mink is a terrestrial animal which prefers wet lands—marshes, muskeg, riverbanks, and lakesides. It burrows just above water level into riverbanks and also uses beaver lodges and muskrat houses, and dens among rocks and under tree roots, where it makes a nest of grass and feathers. It is almost totally carnivorous, quick enough to catch fish but relying more on amphibians, rodents, and aquatic birds, although it may eat berries in the autumn. Its soft and luxurious pelt has been coveted by humans for many years, and the great commercial demand led to large-scale trapping and eventually to fur farming, which has resulted in its domestication. It is a very agile and versatile animal, which climbs and swims and is a good jumper. It is mainly nocturnal, solitary, and territorial and is very aggressive to conspecifics except at breeding time.

Mink have a five-year breeding life and produce their kits in March after a gestation period of only 40 days, although this may be extended by 30 days through delayed implantation. A female may have up to 10 kits, although a litter of five is normal. The babies are weaned when they are six to eight weeks old and transferred to separate raising pens, where they moult in the fall and grow their winter coats. On fur farms, future breeders are then selected, and the rest are killed. The mink's natural color is rich chocolate brown to black, usually with a few white spots on the chin, but farmers have selectively bred a range of mutants. It has been genetically determined that mink could be produced in 200 color variants, but only a few of these are commercially farmed. Black mink, first bred in the world on North American farms, are highly coveted. Other favored colors are white, mahogany, sapphire, pearl, and pastel. In 2005 the farm population of 642,000 female mink in the United States produced a crop of 2,628,000 pelts. It takes approximately 50 mink pelts to make a full-length fur coat

Foxes (*Canidae*)

Fox farming is believed to have commenced in Alaska in 1885, followed by Canada in 1892, from where animals were supplied to other potential breeders in North America and in Russia. They are now one of the mainstays of the fur farming industry, second only to mink in numbers kept and in the annual crop, with the advantage that it takes only 20 fox pelts to make a full-length fur coat. China is currently the world's major producer of fox skins, marketing 3.5 million in 2005, followed by Finland with 2.2 million. Two species of foxes are involved in fur farming, the red fox and the Arctic fox. The red fox (*Vulpes vulpes*) is a wide-ranging and adaptable species, found throughout the Northern Hemisphere in both the Old and New Worlds, but even across this vast range it is considered a single species. Its coat is an unmistakable reddish brown, with white belly and black legs and feet, plus the distinctive reddish-black bushy tail with a white tip. Adult foxes vary in size, with weights up to 17 pounds (7.7 kg) recorded in Europe and 11 pounds (5 kg) in North America. The red fox favors both forest and open country, and although nocturnal, is often seen abroad during the day and has invaded suburbia, living beneath houses and raiding the garbage at night. There are several natural color phases of the red fox. The silver fox is a common color variant, with a black coat and white-tipped guard hairs, giving it a silvery or almost black color, and its tail is tipped with white. The cross fox is a reddish-brown morph with a black dorsal stripe and a stripe across its shoulders, forming a cross. The black fox has black-tipped guard hairs. These color phases may occur within the same litter, but are more prevalent in colder, northern regions. In fact, they rarely occur in the southern parts of the fox's range. These naturally occurring mutants gave fox farmers a good start when they began selectively breeding foxes to produce the range of colors now available. In addition, fox pelts may be dyed, and an even wider range of color options has been provided by cross-breeding the red fox with the Arctic fox (*Alopex lagopus*). The latter has two naturally occurring color morphs: blue, which is chocolate brown in summer and light brown with a blue tinge in winter; and white, which is gray to brownish gray in summer and white in winter. In continental North America (on the tundra north of the tree line) only 1 percent of Arctic foxes are blue, whereas in Greenland half the foxes are blue, and in Iceland most are blue. Blue Arctic foxes have produced kits when artificially inseminated with red fox sperm.

Giant Cane Rat (*Thryonomys swinderianus*)

The giant cane rat or grasscutter rat is a very common herbivorous animal of the grasslands and forest clearings of central Africa, from Gambia east to Sudan and south to Namibia. It has a coarse-haired, grayish-brown coat, a large head, and a long, lightly haired, rat-like tail, and it could be mistaken for a coypu or nutria. When mature, male cane rats weigh about 17 pounds (8 kg) and females 11 pounds (5 kg), and are favorite bush meat, fresh or smoked, that bring higher prices in the markets than beef. The cane rat is not a burrower, but may make use of the holes made by other animals, or it may rest among rocks or hide in tall grass. Young cane rats are precocial, are born fully furred with their eyes open, and are mobile soon after birth. The normal litter of four, born after a gestation period of about five months, has been doubled with good feeding in captivity, and they reach sexual maturity when about one year old.

True to its alternate name, this rodent cuts grass, chewing through the stems to cut them down, eating the succulent internodes, and leaving the rest. It is a very wasteful way of eating, and a vulnerable one too, as the scattered stems draw hunter's attention to the rats' whereabouts. It has been farmed in West Africa for at least three decades, its intensive breeding having been pioneered by Ghana's Wildlife Department in the 1970s; and it is one of the projects assisted by The World Bank. It has also been farmed in Benin and Togo, and rats have now been made available through agricultural extension services to Cameroon, Gabon, Senegal, Cote d'Ivoire, Nigeria, and Ghana. Attempts have been made in recent years to improve the cane rat stocks by selecting for adaptability to a life in confinement. They are easy to care for as they are totally vegetarian, and captive animals are fed fresh-cut grass, cassava, and sugar cane.

Unfortunately, cane rat farming is experiencing problems. Unlike most rodents, their reproductive rate is very slow, and it takes well over a year from conception to the growth of the young to marketable age. Also, although they are usually social in their behavior, pregnant females must be kept alone or males may cannibalize the young, which complicates their housing. Then, during the dry season, the farmers have difficulty finding sufficient fresh grass for them. Consequently, some farmers have already given up as they consider the return for over a year's work is not worth the effort compared to going into the forest and shooting or snaring a duiker or a mangabey, while those animals last.

7 Conservation by Domestication

This chapter is concerned with zoo mammals, but only those species that can still be considered the prerogative of zoos, whether they are institutional or privately operated. This includes mostly the larger, charismatic creatures such as lions, tigers, gorillas, orangutans, giraffes, and rhinos, all still typical zoo animals. In contrast, representatives of the other classes of vertebrates, even the largest species such as ostriches, crocodiles, and giant snakes, plus many of the smaller mammals, have in recent years become the stock in trade of pet breeders and wildlife farmers. Within living memory zoos were consumers of wildlife, but the modern zoo is now a producer, especially of endangered species.

Until well into the second half of the last century, zoos were not major centers of wild animal reproduction. When I entered the profession in the mid 1950s, the main objective was visitor entertainment, despite the scientific aspects of some zoos. Chimpanzees had tea parties on the lawn, feeding time at the lion house was advertised as a daily spectacle, and gorillas and many other social animals spent their long lives in solitary confinement. Zoos were rivals in acquiring rare or unusual animals, and were able to replace their losses from the wild. In the 1960s marmosets could still be purchased by the dozen from importers, and expatriates around the world regularly sent animals, mostly single ex-pets, back to their home-town zoo. In the past four decades zoos have changed, fortunately, in image and purpose. They now cooperate in the breeding of rare species, many of which are considered global rather than individual zoo populations, and their breeding is organized in the species' best interests.

The outcome of the great advances made in the past few decades is that most zoo mammals are now captive-bred but in conditions that are far from natural, and domestication is therefore unavoidable. Physically reflecting the modernization of zoos, numerous multimillion dollar replicas of the Amazonian rain forest, the East African savannahs, and other biomes now grace zoos around the world. They

Western Lowland Gorilla   *Exports of wild-caught lowland gorillas from their countries of origin now rarely occur, as even rescued orphans may be fostered in orphanages in their native lands, with a view to reintroduction. The zoo gorilla population is therefore totally dependent for its perpetuation on its breeding success. Although the birth of the first zoo gorilla did not occur until 1956, over 1,000 babies have since been born.*
Photo: Courtesy Arpingstone, Wikipedia.com

provide more interest for the animals, in some cases allow a certain amount of natural foraging, and certainly enhance the visitor's experience. But there are still no factors in their lives that will continue the process of change that resulted in their evolution to the point when their domestication began.

The possibility of zoo animals becoming domesticated was recognized early in the nineteenth century by Sir Stamford Raffles, founder of Singapore and cofounder of the London Zoo. When the zoo opened in 1825, he described it as an "establishment where animals could be kept for the purposes of domestication." Fifty years ago Professor Heini Hediger, one of the twentieth century's most distinguished zoologists, said that "animals in zoos must be protected from the effects of domestication, and the zoo biologist must watch for its early signs." Domestication is indeed happening, for animals that breed regularly under human care must eventually become domesticated. The great advances in zoo animal husbandry in the latter half of the last century have hastened the process and involved numerous species, including many unlikely ones. The improvements in animal care increased reproduction and thus domestication, whereas the previous ease of restocking from the wild did not encourage regular breeding. Even species like the lowland gorilla, considered for many years an unlikely breeding prospect, finally bred for the first time in 1956 at the Columbus Zoo, where 30 others have since been born. Similar successes in other zoos have resulted in practically all zoo gorillas now being captive-born.

The purpose of zoos has always been to display wild animals, exotic or native, and only domesticated exotic species were generally kept. Thus, long-domesticated animals like camels, llamas, and yak were acceptable zoo animals, but Holstein cattle, Merino sheep, and Shetland ponies were relegated to the children's zoo, if they were kept at all. Now, many typical zoo animals, such as the Barbary sheep and mouflon, easily bred species with little or no new blood since the original imports were made many years ago, are undoubtedly domesticated. But many other zoo mammals have been maintained for numerous generations and must also be considered on route to domestication. Some, like the Przewalski horses, Pere David's deer, Arabian oryx, and European bison, in which the ancestors of all the living individuals were captive-bred, are obviously already in the early stages of domestication. All four species were saved by captive breeding, but they are now thriving in protected and controlled areas where mate selection occurs naturally, and limited natural predation for some. Leopards, tigers, lions, and jaguars have been breeding in zoos for many years and for many generations. Chimpanzees, orangutans, gibbons, and many lesser primates, including Barbary apes, mandrills, and lion-tailed macaques, have similarly been maintained for many generations. Numerous antelopes, deer, zebras, hippos, and rhinos are bred annually in zoos. Regularly bred rodents include the Patagonian cavy or mara, the capybara, and the large crested porcupines. Many of these animals are not new zoo breeders, some having reproduced for the first time over a century ago, such as the ring-tailed lemur and Burchell's zebra at London Zoo in 1858 and the Brazilian tapir at Hamburg in 1868. But most of those early breedings were isolated occurrences, not the sustained breeding that has now occurred for many generations, and as a result of which domestication is unavoidable.

The primary objective of modern zoos and related institutions is the conservation of species, which can only be achieved through breeding. Captive breeding is considered an essential component of wildlife conservation, acknowledged by the World Conservation Union and the World Wildlife Fund. Conservation in the zoo—"ex situ" as that is now called, as opposed to "in situ" in the animal's native habitat—is synonymous with domestication. It is unavoidable for all zoo species that continually reproduce. However, unlike the commercial breeders of wild animals, zoos are attempting to maintain the purity of their stock and therefore characteristic species, and are now also very careful about inbreeding and its potentially harmful effects.

The behavior of wild animals evolved in response to conditions in their natural environment, with natural selection producing adaptations over many generations to exploit the available resources of food, water, and shelter, while at the same time avoiding their predators and thus ensuring their survival and evolution. Captivity changed all that. The captive environment restricts their movement, and provides their food and shelter. Zoo animals cannot select their own mates, protect their territory from potential usurpers, or do ritual battle with conspecifics to possess and control a harem. Boredom is a continual problem, alleviated somewhat by naturalistic displays and behavioral engineering that allow animals to live more natural lives. Social groupings that encourage interaction, activity, and of course breeding, rather than the solitary confinement of old, are also a great improvement, but the zoo environment obviously still falls far short of the natural one. The pressures are different, and it is these pressures that have influenced the direction of the zoo animal's domestication and produced changes in their genetic constitution.

They exhibit changes in their behavior, involving psychological changes (in their mental or emotional processes). They may also change physiologically (in their natural biological processes) and even anatomically (in their structure). These dangerous side effects of domestication cannot be totally avoided, but some may be reduced by concerned husbandry.

There are only two long-term options for zoo animals, at least the rarer species. Their total extinction can be postponed, possibly forever, through their captive maintenance and control, while in the outside world their ancestors are lost. As time goes by they will be so changed they would be unlikely to survive in the wild even if habitat is available for their reintroduction. Or they can be released into a suitable habitat, even semicontrolled areas like those the plains bison and the Pere David's deer now occupy, with the original threats to their survival removed. Either way, they must be kept as close as genetically possible to their ancestral forms.

An animal's behavior either is genetically induced or learned in the early stages of life, or is a combination of both genetics and learning. The zoo animal's environment differs totally from that of the wild animal and has a major effect upon its behavior. The major differences concern freedom of movement, food, social organization, and breeding, especially artificial selection. These aspects of control all have a serious impact upon the animal's lifestyle and eventually produce changes to their phenotype (their observable appearance) and their genotype (their genetic makeup). The loss of freedom is the most criticized aspect of zoos; judgment is rarely passed on the manner of their feeding or breeding, and their social welfare is only a concern when it involves life in solitary confinement.

■ CONFINEMENT

The greatest change from a wild existence to captive life is the loss of freedom, and confinement is one of the three most potentially damaging aspects of zoo animal husbandry, with diet and breeding having equally important consequences. Habitats are occupied by species, and an animal's own particular area within the habitat is known as a territory. Although it is now known that most wild mammals are not necessarily free to wander over a large area, the territories of the major vertebrate animals are very much larger than could ever reasonably be provided in the zoo. It is true, however, that many captive animals do not use all their space. The lion is content to rest in the shade after eating, but its enclosure has become its territory, and it will attack intruders, humans included, who enter it. Within their animal's restricted territories, zoos must provide an environment that allows them to undertake their other natural daily activities, including climbing, swimming, burrowing, sleeping, sunbathing, wallowing, avoidance of their companions if they wish, and protection from the elements, as appropriate for the species. Zoo animals do not need space to forage for food as they would in the wild—where the food supply determines territory size and population density. When food is provided, animals lose their urge to travel great distances in search of it. Confinement reduces the ability to be active, while the regular provision of food reduces the need to be active. Combined, these aspects of control dull the senses and affect musculature and physiology. At least now that most zoo animals are captive-born, the sudden massive change from freedom to close confinement does not exist, and they grow up knowing only the behavior of their parents and perhaps other conspecifics, and their own restricted environment.

Confinement is unquestionably the aspect of zoos that results in most behavioral changes—including psychologically, socially, territorially, and acceptance of the close proximity of man, which is achieved through taming. Wild animals have specific flight and fight distances that vary according to the species, which they must maintain in order to feel safe from a perceived threat. For the truly wild animal unaccustomed to man, the zoo enclosure rarely provides this space, and to avoid animals living in a state of constant fear or injuring themselves when they flee, their flight distances must be reduced. Taming—reduction of the flight-or-fight reflex— also aids domestication. It allows close contact, handling, investigation, treatment, the removal of young for inspection or for supplemental feeding, and animal movement between pens or zoos with less risk to the animals and staff. The keepers may be regarded as members of the animal's social group. The major disadvantage, other than aiding the process of domestication, is that it renders animals unsuitable for reintroduction. Taming is not necessarily synonymous with tameness. The caged tiger's acceptance of contact with its keeper through the bars does not mean that it will allow safe entry into its cage.

The natural social behavior of the wild animal, whether a solitary or social species, is totally disrupted by long-term confinement. Solitary polygynous species that meet just for mating (such as tigers and polar bears), after which the male then seeks another mate, are frequently kept together in zoos. Polar bears have even been kept in groups. When a pair of normally solitary creatures are prepared to live compatibly, they are generally allowed to do so, as this avoids the risk of serious

Snow Leopard *Threatened mainly by hunting for its luxurious pelt, and with only about 5,000 animals now surviving in the wild, the snow leopard is a managed captive species, with a population of almost 700 animals in the world's zoos.*
Photo: Photos.com

confrontation when introducing strangers for mating. Some may be kept singly, the giant panda for example, but aggression or disinterest towards their unnaturally chosen mates, during the very short window of opportunity for mating, has been largely solved through artificial insemination. Within the animals that live in organized groups, a specific behavior has evolved to maintain relative peace in the group. It is based on precedence or ranking and is known as the social or dominance hierarchy or simply the "peck order." All the members occupy a position within the order and are led by a dominant animal which may be male (in zebras, bison, and guanaco) or female (red deer, African elephant, and wild ass). Subordinate animals know their place and behave accordingly. Social ranking involves strictly observed behavioral and ceremonial protocols, and is maintained by rebukes from the dominant members and by ritual fighting, such as the head-butting clashes of the wild sheep and goats, to maintain their position. Zoo groups of these animals are rarely large enough to contain several adult males that can interact in this manner. In many species in the wild, the youngsters disperse at a set age, to find their own mates and establish their own territories, or to avoid the wrath of the herd leader. In the zoo, the curator decides when they should depart.

Zoo hierarchies may be disrupted by the frequent movement of animals, out at least, to prevent overcrowding, whereas introducing new members to an established social group is virtually impossible. Life within a zoo group in a small enclosure provides problems for prospective mothers who seek a secluded place to give birth and then hide their baby. In social groups containing several males, such as baboons, subordinate males may not be able to hide, to avoid posing a threat to the dominant male. Even providing inadequate food dishes that allow senior animals to commandeer the feeding station and select the choicest items can be stressful, and of course unhealthy, for subordinate animals. This can be solved by feeding a complete diet, so that all the leftovers are the same, but such diets may cause even worse problems.

■ DIET

The search for food is the dominant activity of all wild animals, and selective pressures to improve their efficiency in acquiring and consuming their food, and obtaining full benefit from its nutrients, are constantly being improved. Food acquisition is an almost continuous task for most wild animals during their waking hours. Insectivores like the giant anteater spend the day searching the savannah for ant and termite nests and gathering the insects on their sticky tongues. Bands of monkeys forage almost continuously for buds, flowers, and fruit as they move through the treetops. The grazing and browsing herbivores generally have an easier time as grass and leaves are usually plentiful items, but eating is still an almost continuous act, interspersed with periods of resting to ruminate. In contrast, the carnivores (excluding the vegetarian giant panda) take considerably longer to locate their prey than they do to eat it. Hunters like the stoat and marten may spend hours searching for a fresh rabbit trail or locating a gray squirrel, and the large cats are certainly not successful every time they stalk and chase a potential victim.

The zoo environment removes the need for animals to search for their food. For the omnivores, a combination of plant and animal foods is provided in a dish. This leaves a lot of free hours, but behavioral engineering—providing many feeding or foraging stations where animals have to find their food—has helped to reduce their boredom. The herbivorous ungulates are provided with browse (in the form of leafy branches) and grass (usually in the form of baled hay), as they are unable to browse naturally and many zoo enclosures do not permit grazing. They also have supplements (pellets high in protein and vitamins). However, eating out of a hay rack or a food trough does not simulate natural browsing or grazing, psychologically (as it does not allow the selection of plant life), physiologically (due to the nature of the food, which may have too much or too little fiber), and perhaps not even nutritionally (due to the unsuitable composition of the food). But at least they are occupied, usually for hours, chewing their fibrous diet.

Some aspects of the wild carnivore's natural feeding behavior cannot be reproduced in the zoo. Large carnivores cannot be allowed to kill their own prey, and even the smaller species are rarely given live rats or mice. The larger predator's stalk or ambush, followed by the explosive burst of energy of the chase and kill, can never be repeated in the zoo. Even the more natural feeding policy of placing the carcass of a cow or pig in the tiger enclosure for them to eat their fill, and then cover the remains for a few days before gorging again, would never be allowed in a zoo open to the public. The closest public zoos have come to a natural feeding regimen is the implementation of a starvation day once weekly, as big cats in the wild generally do not make a kill every day.

Even more important than the acquisition of their food, which has behavioral consequences, are the actual nutritional aspects. Unsuitable diets have potentially serious consequences that may affect the animal's anatomy and physiology. Few zoo animals can be fed naturally, and the suitability of their "replacement diets" has been the subject of considerable debate over half a century. Animals must have a nutritionally complete diet, which some zoo staff believe can be provided only through the use of prepared foods such as manufactured "pellets" or "cake," every bite of which is nutritionally complete with all the nutrients supposedly required by the zoo animal, although the actual requirements of many species are still unknown. Such simplified diets may be adequate for the rapid growth of the domesticated hog destined for the slaughter house at the age of six months, or for increasing the Holstein's milk yield; but for zoo animals, diet means more than simply providing adequate nutrition.

The use of manufactured complete diets has already resulted in unwelcome changes. Muscles degenerate when they are not used, and human control has already affected the size and shape of the maxillary muscles of captive lions that received fortified hamburger all their lives. Pellets do not provide adequate wear for the molars or work for the jaw muscles, and when moistened in the stomach they turn to mush and pass through the intestine quickly, without allowing sufficient digestion time. It is also well known that unsuitable diets, especially the lack of fiber, can effect changes in bird's digestive tracts, and will no doubt eventually have the same results for mammals. Lack of fiber is fatal to the tree porcupine, whose gut collapses without the required bulk provided by tree bark. The most surprising recent discovery, however, was the fact that excess fiber in the diet can produce just as dramatic changes as a deficiency. There was considerable modification of the giraffe's digestive system after receiving a diet of lucerne hay for years, a forage until then considered most suitable for the leaf-browsing mammals. The problem of such physiological changes due to unsuitable diets will have serious consequences for animals eventually reintroduced into their natural habitats. Major changes to an animal in its lifetime, and then to succeeding generations, whether behavioral or physiological, are not totally inheritable by the next generation, but in time produce the familiar effects of domestication. Long-term domestication is known to produce changes to skulls and teeth, and the bones of dogs found at prehistoric sites are identified from those of wolves by their smaller and shorter skulls and smaller teeth. Long-domesticated dogs and ferrets have longer digestive tracts than their ancestors, reflecting their increased consumption of carbohydrates, requiring more digestion time than protein in the carnivore's simple gut.

Other zoo professionals believe that more natural diets, based upon foods resembling the wild animal's intake, are more suitable for their animals. These provide more variety and interest and therefore combat boredom. They also provide occupation (more selection and chewing) and have healthier consequences for the animal's teeth, jaw muscles, and intestinal tract. They are certainly more stimulating. Years ago in the London Zoo's Lion House, the big cats became highly excited as I approached with their meaty joints at feeding time; nowadays tigers and lions regard their lump of "sausage meat" with disinterest. A midway approach, providing a combination of nutritionally complete items with more natural foods, is considered by many the best way to stem the changes that will undoubtedly result from unsuitable long-term diets. The formulated foods provide the animal's nutritional needs, while the whole foods are occupational and reduce boredom. They have anatomical value also. Chewing—to get meat off the bone as opposed to masticating their food—is considered essential for the carnivore's jaws and teeth. Roughage, in the form of rabbits and chickens for larger carnivores and rats and mice for the smaller species, provides bulk for the intestine. Another major concern is posed by the potential changes to zoo animals' intestinal microflora due to medication and the frequent use, through the lifetime of some species, of prophylactic feed supplements containing antibiotics.

■ REPRODUCTION

Breeding is obviously the most important aspect of zoo conservation husbandry, at least for the rarer species, for nowadays producing common ones causes disposal problems. It is also essential for domestication, which could not otherwise proceed, and the effects of captive breeding on wild animals ranks equally with the changes resulting from the other major aspects of their control—confinement and diet. The main difference between the wild animal and caged animal is in the selection of their mates. Animal breeders replace natural mate selection with unnatural selection, intentionally or accidentally, for certain characteristics in their animals. They favor and therefore select traits that improve their husbandry capabilities, and domestication is therefore aided by the less rigorous selection processes that apply to the captive animal. Continued selection of animals that are more amenable to captive life—are more tractable, readily accept replacement diets, and, most important, are willing to breed—are obviously the animals the breeder will keep. Breeders also raise animals that would not survive in the wild. Baby monkeys abandoned by their mothers are bottle raised. Young northern ungulates born unseasonably late, which in the wild may not have sufficient time to prepare physically for winter, are pampered and saved in the zoo. But these humane practices may also have conservation as well as moral value—for example, when a giant panda mother rejects one of her twins.

Several aspects of captivity affect reproduction. The animal's seasons and cycles can be influenced by a number of factors directly related to their control by humans. A change of environment, such as latitude, temperature, and photoperiod, affect an animal's normal reproductive patterns. When they are kept indoors permanently, in climatically controlled tropical houses, the constant environmental conditions can disrupt their normal breeding cycle. Nutrition can also influence captive breeding, and the consistency of meals in the zoo, contrasting with the fluctuations of supply experienced by many animals in the wild, has enabled seasonal breeders to extend their natural cycle. They grow faster and become sexually mature sooner. They may then be able to breed earlier, at least in the typical herd arrangement of one male and several females, as they are not constrained by a dominant male and do not have to wait until they have achieved the size, antler, or horn growth, or the strength to challenge for control of the herd. Some zoo animals now breed annually instead of every two or three years, as a result of a shorter suckling period and earlier separation of the young from their mothers than would occur in the wild. Young tigers in the wild normally stay with their mother for two to three years to learn how to hunt and survive, but in the zoo they are generally removed when they are weaned at 12 weeks of age, and sometimes even earlier for hand-rearing, which results in their mother's quicker onset of estrus and re-breeding.

Natural mate selection in the wild creates stabilized biological systems that ensure the development of an animal able to adapt to a wide variety of environmental conditions and therefore ensure the continuation of its species. In contrast, artificial selection—when man decides which animals to breed—breaks down these stabilized systems and creates combinations of genes that may not permit the animal's survival in nature. Zoo animals are unable to win mates through dominance, for the typical composition of pairs, trios, or small groups does not allow males to control a herd through the ritual of rutting and fighting. Mate selection in the zoo is almost totally artificial. It may be based on an animal's healthy appearance, age, its past breeding success, acceptance of the captive environment, its tameness, and certainly its compatability, and animals that cannot adapt to a life of close confinement and control are eliminated. Increasing importance is now being placed, however, at least with rarities, on an animal's genetic background, and all other things being equal, the most suitable mate will be the least related individual. Knowing how devastating inbreeding can be to a captive population, every effort is now made to avoid it, except in the few zoos that are perpetuating white tigers and other aberrant animals. Intentional hybridization is now also avoided by most zoos.

Southern White Rhinos   *The second-largest land mammal after the elephant, the southern white rhino is a managed zoo species, with an international studbook and regional breeding consortia controlling its genetics. The large zoo population, of about 750 animals, comprises mainly the offspring of animals officially culled in the 1960s when their national parks in South Africa became overstocked.*
Photo: N. Joy Neish, Shutterstock.com

Hybrids occur when animals are sufficiently closely related for the gametes (the eggs and sperm) to produce a viable embryo. Subspecies of a species can hybridize and have fertile offspring, and closely related species can hybridize and the offspring may be fertile, but it is highly unusual for animals belonging to different genera to produce viable offspring. Hybridization is an onslaught on the purity zoos attempt to maintain in their stock, and entry into studbooks has been refused with regard to animals whose ancestry is unproven, or whose ancestors may have been involved in a mixed-breeding indiscretion years ago. Maintaining pure species, and even pure subspecies, is considered very important in the modern zoological garden, and is aided nowadays by species' survival plans and studbooks for the rarer species. Hybridization may seem relatively easy to avoid, considering it can only occur through cross-breeding animals, but it was actually popular in some zoos in the last century when lions and tigers were hybridized to produce tigons (tiger/lioness) and ligers (lion/tigress), but such intentional crossing is now rare. It can still occur accidentally, however, even in the best zoos. In 2006 a lonely male babirusa (a wild pig from Sulawesi) in the Copenhagen Zoo was given two domestic sows as companions, one of which surprised the zoo world by giving birth to two hybrid offspring. In theory this should not have happened as the two are so genetically remote, even in different genera. However, there are many "generic" animals in zoos, the results of past mixing of the blood. This may have been due to misidentification, the inability to acquire the right mate, or simply because an animal's origins and ancestry were unknown. Consequently, many giraffes, sika deer, wolves, leopards, African lions, pumas, and tigers in zoos nowadays are considered generic. Many tigers are neither Siberian tigers, Sumatran tigers, nor Bengal tigers—simply tigers. Fortunately, there are also many proven pure subspecies of these animals in zoos. Domestication alone does not compromise purity, but intense inbreeding certainly changes animals genotypically, and possibly also phenotypically, producing animals outwardly differing from their ancestors.

Inbreeding or homozygosity is the harmful practice of mating closely related animals, either brother and sister, or father and daughter. It causes a decline of both population and individual genetic variation, reducing individuals that are heterozygous for any pair of genes and increasing those that are homozygous, thus increasing opportunities for mutants to become established. Breeding such closely related animals results in the appearance of more deleterious (injurious or harmful) recessive genes, and reduced fitness, known as inbreeding depression. This affects fertility, successful births, the viability of the young, adequate parental care, and healthy growth, the two most obvious signs being reduced breeding and increased juvenile mortality. The more closely related the animals, the more homozygous genes they will have, affecting their viability; and when animals are continually inbred, most of the offspring will have deleterious traits. Non-inbred populations contain a number of potentially harmful recessive genes that are rendered ineffective by the dominant genes.

Although in the research laboratory strains are not officially considered inbred unless they have a coefficient of almost 100, considerably less inbreeding is cause for concern in the zoo profession, and it is likely that only the Barbary sheep in British zoos have come close to such a high coefficient. Inbreeding is used as a tool in the pet trade to hasten the production of mutants, but is generally avoided in the zoo, except when producing white tigers, white lions, and other mutants. Although most zoos now attempt to preserve their animal's genetic integrity, some inbreeding may be unavoidable—when, for instance. a captive population originates from just a few founders. This has happened with Speke's gazelle (*Gazella spekei*), the Nilgiri tahr (*Hemitragus hylocrius*), and the Persian fallow deer (*Dama d. mesopotamica*), due to their rarity or poor availability.

The dangers of inbreeding have been known for almost a century in domesticated animals, and inbreeding depression in zoo animals was documented in several studies in the 1970s and 1980s. An example of its harmful effects occurred in a herd of dorcas gazelle (*Gazella dorcas*) at Washington, DC's National Zoo. Small antelope from the deserts of North Africa and southwestern Asia, the zoo's herd began with just one pair of probably related animals in 1960, and the later introduction of three others, one possibly related to the first pair. The classic study by Ralls, Brugger, and Glick showed that calves from the most inbred specimens suffered a higher loss rate (from infection and poor viability) in the first six months than offspring from less inbred parents, and the survivors matured slowly, with females reaching breeding age one year later than normal. The mortality rate for the inbred babies was 31 percent higher than from out-crossed parents. The dorcas gazelle is a fairly common zoo species, with opportunities for out-crossing, whereas the Przewalski horse (*Equus caballus przewalskii*) is one of the very rare species affected by inbreeding. Its history has been recorded for much of its captive life, certainly for longer than most zoo animals. Inbreeding has occurred mostly as a result of a shortage of fertile stallions, and resulted in reduced breeding and poor survival rate of the foals. Outbreeding enhancement is the name given to the introduction of genes from a different population of animals to reverse inbreeding depression, as these animals will have different deleterious traits, and their breeding will not result in homozygosity. Unfortunately this is impossible for the horse and other rare species with very low populations.

The simplest way to avoid inbreeding is to mate the least related animals each time, assuming there are sufficient numbers to do this, but the number of individuals needed to prevent inbreeding depends upon the lethal or deleterious genes carried by the parents. The golden hamster is an example of an enormous world population descended from just a few founders, but the viability of other captive mammals has suffered from inbreeding, including the Pere David's deer and the eland. Ideally, to conserve a population's gene pool and prevent inbreeding and its subsequent depression, a population needs at least 100 unrelated individuals, preferably in an equal sex ratio and kept in pairs. This pairing arrangement conserves almost double the genetic diversity than keeping animals in the typical zoo grouping of a male and several females. But zoos individually do not have 100 unrelated animals of a single species, so species survival programs and breeding consortia have been established to provide the numbers needed to maintain genetic diversity, although for many species even the total captive population is nowhere near the required number.

Small populations are more susceptible to disease and natural environmental disasters and are also more vulnerable to genetic problems. They risk the loss of diversity through genetic drift, a random process in which not all of the genes are passed from parent to offspring and are then lost—which is particularly harmful to small populations. There is also the risk of a population bottleneck, an event in which a large part of a species population dies or cannot reproduce, which increases genetic drift and, of course, increases inbreeding as there are fewer animals left in the gene pool.

In a closed captive population, the gene pool is totally dependent upon the genes of the original members—the founders. Animals receive their genes equally from both parents in the form of a copy of one of each pair from each parent. The selection of one allele over the other occurs purely by chance, so problems can occur when breeding involves just a few animals. In the wild, loss of habitat results in small, fragmented populations that are more susceptible to disease and natural environmental disasters and are more vulnerable to genetic problems. Lost genes can only be recovered through the very slow process of mutating, or by breeding with other members of the population, which is impossible when initially small populations, like the giant panda's, have fragmented.

■ SOME OF THE SPECIES

Tigers (*Panthera tigris*)

Tigers have always been the most charismatic of the large cats, especially the Siberian tiger—the largest cat. They have been zoo animals for many years, at least since the eighteenth century when they were kept in London's Tower Menagerie. The London Zoo received its first tigers in 1829, four years after it opened, and captive Bengal tigers bred in India in 1880. Tigers have suffered heavily in the wild, from hunting for sport or for their products, conflict with humans, and loss of habitat, especially in the last century, and only five of the recent eight subspecies survive. They are the Siberian, Indo-China, South China, Sumatran, and Bengal tigers. However, the Sumatran tiger was recently considered a genetically distinct species, resulting from its isolation since the island separated from Malaysia when sea levels rose about 10,000 years ago. The other three recent subspecies—the Caspian, Java, and Bali tigers—were all exterminated in the last century. In the wild, the Bengal tiger is the most plentiful race, with a population of about 4,000 animals in protected areas. There are believed to be about 1,500 wild Indo-China tigers left, but the Sumatran and Siberian tigers are down to 400 individuals each. China claims there are 50 South China tigers in the wild, yet none have been seen for 10 years. Fortunately there are also 50 in captivity, although they are all in Chinese zoos and are said to be descended from six wild-caught animals. The Siberian tiger's long-term prospects are more encouraging, as it has a zoo population of 500, and the Bengal tiger is in a similar position with almost as many captives. But only 60 Indo-China tigers are included in the studbook, and although the Sumatran tiger's zoo population is 235 animals, it is considered underrepresented, meaning it has too many closely related animals in breeding programs. In addition to the pure tigers, there are many generic animals of mixed blood, or of uncertain origin, and numerous white tigers, which are the most controversial of all zoo animals.

White tigers are not albinos; they have light blue eyes, a pink nose, and chocolate-brown stripes on a creamy coat. They were initially genetic mutants of the Bengal tiger, and are unknown in the other subspecies of tigers. Their color results from a double recessive gene, which occurs very rarely in the wild—the last recorded instance being in 1951, a member of a litter captured in India by the Maharaja of Rewa. Called Mohan, he was mated with his daughter, as white cubs are only conceived when the gene for white coloration is inherited from both parents, so they can only be perpetuated by repeated inbreeding, such as father to daughter or mother to son, which is unlikely to occur in the wild. The resultant white cubs produced in India formed the foundation of all the current white tigers on display in zoos and circuses. White cubs born in India after further inbreeding were acquired by the National Zoo in Washington, DC; and the Cincinnati Zoo then also continued their production through inbreeding.

Several of the world's major zoological gardens now exhibit white tigers, and they are favored circus animals. Although they began as mutant Bengal tigers, they have been hybridized with Siberian tigers and with "generic" tigers of unknown ancestry, so many are now subspecifically impure or at least are of uncertain lineage. Their continued inbreeding has resulted in a high inbreeding coefficient and the characteristic problems that include poor viability of the young. Their production has been condemned by the Tiger SSP, which has designed breeding strategies to minimize inbreeding and preserve the genetic integrity of small captive populations of tigers, and dismisses as groundless claims that perpetuating white tigers has conservation value. The production of white tigers is considered by most to be contrary to the objectives of the modern zoo, as their perpetuation depends entirely upon intense inbreeding. Consequently, most zoos shun the animals; but a few of the world's major zoological institutions, which are involved in the conservation of many endangered species, have seemingly been influenced by the increased attendance revenues generated by the mutant tigers.

Przewalski Horse (*Equus caballus przewalskii*)

The Przewalski horse is one of the species, like the Pere David's deer and the Arabian oryx, that were exterminated in the wild and survived only in captivity until animals were recently returned to their native range. It is a stocky horse, beige-brown or dun in color with a dark tail and a stiff, upright black mane. An adult stallion weighs up to 750 pounds (340 kg) and is about 13 hands (52 inches) high at the withers—the base of the neck. It has 66 chromosomes, whereas the domesticated horse has 64, and hybrid offspring have 65 and are fertile. The "P. horse" as it is generally known in the zoo profession, lived in the dry, sandy grasslands of central Asia, where a small population in southwestern Mongolia was discovered in 1881 by Colonel Nicolai Przewalski. At the time all the true wild horses were believed extinct; the last tarpan, ancestor of the domestic horse, dying in 1875.

All the living P. horses stem from several large captures of foals made in 1901–3 by the animal dealer and zoo-builder Carl Hagenbeck. Many of these died on the journey from Mongolia to Europe, and the survivors were divided among several zoos and the Duke of Bedford's estate at Woburn Abbey in England. Thirteen horses from these original animals were the founders of the current population of 1,500, which includes those returned to Mongolia recently. By 1960 the horse was extinct in the wild.

Initially there was little or no exchange of horses between zoos, and the subsequent inbreeding resulted in high foal mortality, congenital defects, and shortened life spans. The degree of inbreeding is measured by the inbreeding coefficient, often expressed as the F number. This ranges from 0 percent in the base population of heterozygous animals to 100 percent in which the animals are fully homozygous—having common ancestors. With the population down to 300 horses in the early 1970s, and with several animals having an inbreeding coefficient of 50 percent and many others having F numbers in the 30 to 40 percent range, it was clear they needed to live in larger herds where they could select their own mates. This led to the formation in 1977 of The Foundation for the Preservation and Protection of the Przewalski Horse in Holland, with the intention of reintroducing horses to their natural range. Initially potential horses for release were kept on five reserves in Holland and one in Germany, from which suitable animals were selected. In 1992, 1994, and 1996, 16 animals on each occasion were returned to the 24,000-acre (9,700 ha) Hustai National Park in Mongolia, where they live as wild animals (or perhaps feral animals would be more appropriate). They cope with the vagaries of the climate without human assistance, and band together in their natural herd behavior to protect their foals from wolves. They have thrived and had increased to 300 animals by 2005.

Barbary Sheep (*Ammotragus lervia*)

This is a large sandy or reddish-brown sheep, with heavy horns that curve outward and backward and then turn inward. Females have smaller horns and weigh up to 143 pounds (65 kg), whereas the males weigh up to 320 pounds (145 kg). Barbary sheep are impressive animals, with heavy throat manes extending down their necks onto the chest and legs. Natives of the mountain ranges of northern Africa from Mauretania and Morocco to the Sudan and possibly Egypt, their habitat includes the old Barbary Coast, from which they derive their name. Active, sure-footed, and very hardy, they are great jumpers, able to clear a typical 6-foot-high (1.8 m) zoo fence with ease, although they rarely do. Despite their name, they are considered to be intermediate between the sheep and goats—not really a sheep, and not quite a goat.

Barbary sheep, or aoudads as they are called by the Berbers of North Africa, have been introduced into southeastern Spain and were liberated in Texas in 1957, where they are now well established and may even be more plentiful than in the wild. They are also very common zoo animals, easy to maintain and breed, and are well known for their ability to thrive for many generations without the introduction of new blood. They live compatibly even with several adult males in the flock, are very resistant to parasites, and tolerant of overcrowding. Their popular displays have graced small, low-budget zoos and wealthy institutional ones, since the modern era of zoo-keeping began. All the Barbary sheep in British zoos are descended from a small flock imported by the London Zoo in 1842, a few years after it opened. Although they must have bred sooner, the first report of births did not appear until 1854, and their continual inbreeding since then has resulted in the British population now being considered homozygous or genetically identical.

There is no doubt that the Barbary sheep is one of the longest-domesticated zoo species, with over a century and a half of captive control and considerable inbreeding. Despite this, they show few signs of deleterious effects, just some loss of size in a few animals and the occasional white foot. When a population drops very low and then increases again, it is said to have experienced a bottleneck that greatly reduces genetic diversity, and such an event results in the lethal genes being removed through the death of the animals. Due to their isolated distribution on well-dispersed Saharan mountain ranges, separated by vast stretches of open desert, there could well have been a previous population bottleneck in the wild flock of Barbary sheep from which the first zoo animals were captured.

Giraffe (*Giraffa camelopardalis*)

Giraffes have been popular zoo animals for many years. Understandably so, for they are the tallest of all land mammals; a mature bull may be 18 feet (5.5 m) high and weigh 3,000 pounds (1,360 kg). There are nine recognized subspecies, with

Reticulated Giraffe   *Zoos began breeding giraffes soon after they first acquired them, in the nineteenth century in London and Paris for example, but large-scale breeding did not begin until the middle of the last century, when many were imported from East Africa after World War II. The large zoo population is now virtually composed of captive-bred animals, and many births occur annually.*
Photo: Scott Sanders, Shutterstock.com

the most plentiful one in zoos being the reticulated giraffe (*G. c. reticulata*) from Kenya and neighboring Ethiopia and Somalia, which has large liver-colored markings starkly outlined with white. Several subspecies are being kept pure, but many zoo giraffes are of unknown subspecific origin or have been hybridized and are therefore simply generic giraffes. Many of the current zoo giraffes are descended from animals exported from East Africa for several years after World War II, to restock European zoos devastated by the war. But their captive history and breeding goes back much farther. The London Zoo obtained some soon after its opening in 1825 and recorded the first birth in 1841. The Dublin Zoo, third oldest of the "modern" zoos after Jardin des Plantes and London, received its first giraffe in 1844 and bred them in 1857. The largest zoo herd of reticulated giraffes lives at Colorado Springs' Cheyenne Mountain Zoo, where 185 calves have been born since 1954.

Giraffes are ruminants, members of the suborder *Ruminantia*, which are animals with a compartmented stomach in which plant fiber is broken down by microorganisms. The ruminants are grouped according to their food selection. The main groups are either grazers of highly fibrous grasses, or browsers of less fibrous leaves, or they are intermediate between the two, eating both grass and leaves as they are seasonally available. The ruminant digestive system has evolved to cope with this varying amount of fiber, with a complicated stomach divided into four sections—the rumen, reticulum, omasum, and abomasum. Giraffes are browsers that select the more easily digested shoots and young leaves of the acacia or thorn-trees, the characteristic flat-topped trees of the African savannah and veldt. These leaves have low fiber and high plant cell content, and the giraffe's digestive system reflects this food selection with poor ability to digest plant fiber, in the small size of their fore-stomach or rumen, and in the structure of its lining or mucosa. Food passes more quickly through their system than in the ruminants like the bison that eat coarse, high-fiber grasses. The zoo giraffe's diet is primarily lucerne or alfalfa hay, which has a much higher protein content than grass hay and less fiber in its leaves, but it does have a very fibrous stem.

One of the most detailed studies of the physiological changes that occur when wild animals are subjected to unnatural zoo diets took place in Germany in the early 1970s, when researchers autopsied two giraffes that had lived in zoos for 18 years. The investigations revealed a major change in the size of the fore-stomach between wild and captive animals. In wild giraffes, the capacity of the rumino-reticulum (the two major compartments of the ruminant stomach) is 27 gallons (105 L), whereas in the two zoo animals it was considerably enlarged, having a capacity of 40 gallons (149 L). Also, the mucosa papillae (the nipple-like projections on the stomach wall) were reduced and almost absent in some areas of the zoo animal's stomach compared to wild giraffes, which had nine times the absorptive surface of the rumen to cope with the rapid fermentation of their leafy diet. These major physiological changes resulted from the excessive fiber in the zoo giraffe's diet, until then considered suitable for browsing mammals.

Southern White Rhino (*Ceratotherium s. simum*)

The second-largest land mammal after the elephant, the southern white rhino is actually a gray animal, its name derived from the Africaans word "weit," meaning wide. This describes the rhino's broad snout and square lip, an adaptation for grazing, unlike the black rhinoceros's hooked upper lip that has evolved for browsing leaves. It is the most plentiful of the African rhinos, with a population in the wild (but only in protected areas) of about 11,000 animals, and it has a large zoo population. Its subspecific relative the northern white rhino (*C. s. cottoni*), is on the verge of extinction, and the other African species, the black rhino (*Diceros bicornis*), numbers only 3,600 in the wild and 250 in zoos. The southern white rhino weighs up to 5,500 pounds (2,495 kg), stands 6 feet (1.8 m) high at the shoulder and is 13 feet

(4 m) long. Its natural range is southern Africa, from northern Cape Province to Zimbabwe, Mozambique, and Angola, where it lives in the savannahs and mopane scrub veldt.

The large zoo population of southern white rhinos stems from the 1960s when the Umfolozi and Hluhluwe reserves in Natal, now called KwaZulu-Natal, in South Africa, became overstocked. Rather than risk overgrazing, and not wishing to cull the animals, many were sent to other protected areas in South Africa, including 87 to Kruger National Park; and numerous southern white rhinos were shipped to zoos overseas, where they have readily reproduced. In 2005 there were 750 southern white rhinos in zoos and similar institutions outside South Africa. The San Diego Wild Animal Park has the greatest breeding record, producing almost 100 calves since the early 1970s, and is followed by Whipsnade Wild Animal Park in England, where over 50 have been born. The records of these animals and their genetic management is overseen by The International Studbook for the Southern White Rhino, maintained by the Berlin Zoological Gardens, and by several regional breeding consortia that operate globally under the auspices of the Species Survival Commission (SSC) of the IUCN. It seems unlikely that zoo rhinos will ever be needed from outside southern Africa for reintroduction into their former range, as the Natal Parks Board is still supplying surplus rhinos to other reserves and game farms from their Hluhluwe-Umfolozi National Park. The captive white rhinos will therefore remain in zoos for a long time, perhaps forever, becoming more domesticated with each generation.

Western Lowland Gorilla (*Gorilla g. gorilla*)

The largest anthropoid ape and therefore the largest primate, the western lowland gorilla is one of two subspecies of the western gorilla (*Gorilla gorilla*). Despite being hunted, as it is favored bushmeat, it is still fairly plentiful in the wild, with a population believed to be over 10,000. It is the most well-represented gorilla in zoos also, with about 750 animals around the world. Its natural distribution is Central Africa, from southern Nigeria east and south into Cameroon, Gabon, Equatorial Guinea, and southwestern Central African Republic. There it lives in family groups, comprising a dominant male and several mature females and their offspring of varying ages. They live in the rain forest, where they are purely vegetarians, and are basically terrestrial, but unlike the other great apes, they sleep at night on the ground on a bed of vegetation.

Adult male gorillas are the most impressive primates. Weighing up to 440 pounds (200 kg), they have a very broad chest, muscular neck and strong feet and hands. The skull is large and wide, and they have a bony ridge above the eye and flared nostrils, and a prognathous jaw in which the mandible (the lower jaw) projects beyond the maxilla (the upper jaw). Their heavy head is accentuated by the large hump on the nape. Males over 12 years old are called silver backs, for they have a dark-gray head and forelimbs, and short, light-silver hair on their backs and rumps. The hair on their forelimbs is long and shaggy. They are long-lived, a span of 50 years being possible, and the famous Philadelphia Zoo gorilla "Massa" reached the age of 54.

Gorilla displays in zoos have improved tremendously from the small barred cages of old, and most now live in expensive naturalistic displays, usually in social groups. Surprisingly, they were one of the last of the regularly kept zoo primates to breed. It happened for the first time in the Columbus Zoo in 1956, and then with increasing regularity throughout the world. Over 1,000 captive births have been recorded since then, and most of the present zoo animals are captive-bred. The Columbus Zoo has bred 30 gorillas since the first one, 48 have been born at Cincinnati Zoo, and over 60 at the most famous and largest captive collection of gorillas, founded by the late John Aspinall at Howletts in Kent. Their gestation period is 8½ months, and the mother carries her baby for the first four months. When the baby can cling, it rides on her back, and is carried for at least three years. Wild-born gorillas are unlikely to be available to zoos again, except perhaps for orphans rescued after their mothers have been killed by poachers. But such waifs are increasingly being raised in orphanages in their countries of origin, with a view to their eventual return to the forest. Like elephants, zoos will be totally dependent in future upon their own breeding efforts for maintaining their stocks of gorillas.

Pere David's Deer or Milu (*Elaphurus davidianus*)

Pere David's deer is the most unusual of all deer, both in appearance and in its recent history. Adults are dark reddish brown in summer with a black spinal stripe, changing for the winter into a coarse, dull coat of dark grayish fawn with a soft brown underlayer resembling wool rather than hair. Stags have a shaggy throat mane and deep preorbital glands that are further accentuated by their dark lining. Adult males stand up to 53 inches (1.35 m) at the shoulder and weigh up to 440 pounds (200 kg), the females are 48 inches (1.2 m) high, and weigh up to 360 pounds (163 kg). Their antlers are very unusual, having a long straight tine that points backwards and a main beam that sticks straight up and branches once or twice. The rear shaft is used to plaster their backs with mud. Stags often grow two sets of antlers in one year, and females occasionally have short, unbranched antlers. Their tails are uncharacteristically long and donkey-like, and they have large splayed hooves similar to those of the reindeer. They have a habit of soaking in deep water, whereas the other deer that wallow generally do so in mud.

Natives of China, where they are known as milu, they lived on the coastal flood plains and marshes of the Yangtze and Yellow rivers. They have been extinct in the wild since sometime in the nineteenth century, and survived only in Peking's Royal Hunting Park—where French missionary Pere Armand David saw the few remaining animals in 1865. Western interest in this new species eventually resulted in several deer being shipped to zoos in Europe. Meanwhile, in 1895 the hunting park's walls were breached by floodwaters, allowing most of the deer to escape, to be killed and eaten by the peasants. Then, five years later, troops stationed in Peking during the Boxer rebellion (1899–1901) shot the remaining deer, and the milu was extinct in China. When news of this disaster reached Europe, the zoo directors sent all their milu to the Duke of Bedford's Woburn Abbey estate where they could be kept in a large herd. Of the 18 animals the Duke received, one stag and five females bred and were the founders of the total current world population.

After suffering difficult times during the two world wars, the increased size of the herd allowed animals to be sent to zoos around the world.

After spending untold years in deer parks, the milu was already semidomesticated when the first individuals were shipped to Europe, and they were obviously closely related. Inbreeding has since been a concern due to the small founder population of the species. The effects were especially troubling in zoos, where calf production and survival did not match Woburn Abbey's success, where more natural selection was possible. The Woburn herd averages 300 animals from year to year, and achieved almost double the ratio of births than the zoo animals living in smaller herds—often just pairs or trios, that had been selected artificially. Unfortunately the ability to maintain any ungulate in such large captive herds is very limited. The captive population increased to the stage where animals could be returned to China, and in 1985 22 deer were released in Nan Hai-tsu Park, part of the former imperial park outside Beijing where Pere David first saw the deer. Then in 1987, 39 milu donated by several British collections were released in a larger wetland reserve on the coast at Da Feng, north of Shanghai. These animals have thrived and the population in China now numbers at least 1,300, while in zoos around the world there are about 1,000 animals. Due to the natural breeding of the largest segment of the milu's population, and the impossibility of keeping mating records, the international studbook for the species has been discontinued.

Elephants (*Elephantidae*)

Elephants are not domesticated animals. Despite their long association with man, going back at least four millennia in Asia, they are the classic example of animal taming and training, not domestication, for until recently they simply did not breed often enough. The Asiatic elephant (*Elephas maximas*) is depicted working for man in bas-reliefs from 2000 BC in the ancient Indus civilization of Mohenjo-Daro, and has been used for ceremonial purposes for centuries. They were likely first used for war in 1100 BC and were deployed by the Persians against Alexander the Great at the battle of Gaugamela (331 BC). In recent years the Asian elephant has been a valuable asset in the teak forests of Burma and in carrying visitors in national parks in India and Nepal. But despite its long history of use by man, there is no evidence that it was ever domesticated, merely tamed after being caught as calves in the wild. Working elephants were rarely bred, as their pregnancy and calf-raising effectively removed them from the work force for some time.

African elephants (*Loxodonta africana*) were also tamed and trained for war by the Egyptians, the Carthaginians, and the Numidians, who occupied the region currently called Algeria and Tunisia. In those days elephants, probably a subspecies of the savannah elephant, lived in North Africa, and 37 trained animals played a key role in the Second Punic War (218–202 BC) between Carthage and Rome, assisting Hannibal on his famous march across the Alps. Since the time of the Roman occupation of North Africa, however, until quite recently the African elephant was considered untrainable. This was mainly because the North African elephant had been exterminated by the Romans, and although the elephants of West and Central Africa still survived, the tribal cultures there did involve elephants as they did in Asia.

Asian Elephants   *Used by man for centuries, the Asian elephant has always been considered a domesticated animal, but all the animals used for war, work, and ceremony were wild-caught during elephant roundups, then tamed and trained. Breeding, and thus the start of domestication, did not occur with any frequency until recent times, in zoos and sanctuaries in their native lands.*
Photo: Denise Lafferty, Shutterstock.com

The preservation of elephants in the face of habitat loss and poaching (especially in Africa) is of international concern. In regions where mature bulls have been reduced, poachers have then concentrated on the matriarchs, thus upsetting the herd hierarchy. Their lowered numbers have also created a shortage of elephants for work and ceremonies, with considerable reduction in the numbers captured in Thailand and Burma. The elephant roundups, when wild animals were driven into stockades and young animals were subdued with the help of trained elephants, are occurring less frequently, and they have already ended in Sri Lanka and India for humane reasons.

Fortunately, captive elephants are now breeding, although not with the frequency that is possible or desired; but it is certainly sufficient to finally begin the process of domestication. The first captive birth in the United States occurred in the Cooper and Bailey Circus in 1880. This was followed by other occasional births in the world's zoos, but they were generally accidental affairs, as most elephants were kept in unsuitable conditions, often singly and rarely in the natural social groupings of the species. During the twentieth century, and mostly in its last two decades, about 120 Asian and a dozen African calves were born, a reflection of their improved husbandry and housing.

Until recently, zoos were not happy homes for these large mammals. No other zoo animal is more dependant upon consideration of its social, physical, and psychological needs, and direct human contact is also essential to maintain the health of captive animals, especially their hides and feet. The required space, and cost, of keeping elephants in their natural social arrangement, including an adult bull, were deterrents that have only recently been overcome. More zoos are now keeping them in small herds and are able to cope with bulls. For those that cannot, or have a poor breeder, the increasingly successful use of artificial insemination (AI) avoids the difficult task of either keeping a bull, having an infertile one, or shipping cows to another institution for breeding. In 2000, two African elephants gave birth to healthy calves after being artificially impregnated, and then in 2005 and 2006 the same two cows also produced healthy calves, again by AI. The Portland Zoo is the world leader in elephant production, with 27 births in total since "Packy" was born in 1962. He has since sired seven calves himself, the only second-generation captive breeding group in the world.

In their native lands, elephant sanctuaries and orphanages are also being developed, and calves have already been born there. They exist in Kenya, Malaysia, Thailand, and Sri Lanka, and calves have been born in northern Thailand and in Sri Lanka—where 20 calves have been born since 1984 at the most famous orphanage at Pinnawela. Even though captive elephants live for 50 years or more, the zoo population will dwindle unless the rate of reproduction increases, as imports of calves from their countries of origin become more difficult and will likely soon be impossible. Although zoo elephants are breeding, and can finally begin to qualify for the domestic label they have worn for so long, their current reproductive rate will not maintain their existing captive numbers.

8 Free Again

The world is full of the descendants of domesticated animals that escaped or were released and are living wild and free. They may be back in their countries of origin or on another continent. They may be able to meet up with their own kind still living there, but most are new to their environment and to the animals already occupying it. They include some most unlikely animals. Pigs, whose ancestors came from Europe, root for food on the lower slopes of Hawaii's Mauna Kea, in New Zealand's Te Urewera National Park, and on the banks of Queensland lagoons, where introduced Indonesian water buffaloes wallow. The descendents of domesticated rabbits, ferrets, cats, dogs, sheep, goats, horses, and cows, are free again, living off the land without human assistance. Domesticated amphibians, fish, reptiles, and birds have also made their homes in many new lands. Common carp, farmed for centuries in Asia, are now established in waterways around the world after escaping from their ponds or being deliberately introduced for sport fishing. Twenty-eight species of exotic fish, mainly descendents of aquarium species, now thrive in Florida's fresh waters. Farmed tilapia have been released or escaped and now thrive in waterways far removed from their ancestral African homes. Unwanted pythons and iguanas live in the Everglades, and the clawed toad, first a widely used laboratory animal and then more recently a popular pet, has established free-living populations in several countries including the United States, where it is a pest in California.

Many domesticated birds also thrive in environments far removed from the lands in which they evolved. Flocks of red-crowned parrots, ring-necked parakeets, cockatiels, and budgerigars fly freely over California, and several species of parrots are established in Florida. Domesticated ducks live semi-wild on the ponds of city parks around the world, together with their mallard ancestors and hybrids of many colors. Another domesticate—the South American muscovy duck—also lives in city parks and has well-established, free-living populations in Florida and Texas. It mates with the mallards and their domestic descendents, but the ducklings are infertile. The mandarin duck is established in Great Britain from escapees and deliberate releases, and an estimated 50,000 Canada geese live there also, the descendents of geese introduced as ornamental waterfowl in the seventeenth century. Egyptian geese, escapees from private waterfowl collections, are also established in England and Europe, especially in the Netherlands. Domesticated mute swans have colonized the eastern United States, southern Canada, and the Great Lakes region, after being released in the Hudson Valley in the 1800s. Australian black swans are also established in the wild in Texas, New Zealand, Europe, and Great Britain. Domesticated turkeys live wild and free on Martha's Vineyard, and peafowl thrive in California, Mexico, and New Zealand. Ostriches, imported into South Australia from South Africa in 1881 for farming, were released when the industry collapsed, but survived in the outback. With the recent revival of ostrich farming, some of their descendants were caught and formed the basis of the current Australian ostrich farming industry.

These new settlers are known as feral animals, domesticated species that have reverted to life in the wild, usually not where they evolved. They were able to survive because their natural instincts had only been changed, and not eliminated, during their many years of control by man. To successfully colonize their new homes, feral animals required a supply of food and a reasonable climate. They needed access to suitable mates, and their new environment had to be free of predators and of endemic diseases of which the newcomers had no experience and therefore lacked immunity. They found numerous suitable places around the world, but their successful colonization meant disaster for many of the resident plants and animals.

Many animals domesticated long ago by man, or at least some of their breeds, have shown they can return to the wild despite thousands of years of human control. Predators such as cats and ferrets are perfect examples of animals that have retained their hunting instincts, and their survival hinges solely on the availability of food. Omnivores have even less difficulty finding food. Gallinaceous birds such as the ring-necked pheasant, chicken, peafowl, and guineafowl, scratch for seeds and invertebrates, and their successful introduction hinges more on the local level of predation. Pigs, also omnivores, similarly have an easy time when they are released in a region that supplies food, has a reasonable climate, and lacks large predators. They just have to dig for roots and tubers, raid crops, find a wallow, build a nest for the night, and they are quite happy. It is even easier for the herbivores to resettle far away from their ancestral homes. In the Welsh mountains, the Australian outback, or the American southwest, wherever grass or leaves were available, escapee and released rabbits, deer, goats, sheep, cattle, and horses have survived. However, the ability to revert to the wild does not apply to all breeds of domesticated animals. Some morphological changes resulting from artificial selection would make life in the wild impossible, and breeds with shortened faces, such as the Pekingese and the Persian cat, would never survive on their own.

The wild ancestors of feral domesticated animals have experienced a varied recent history. Some are still very common. Mouflon are still widespread across Europe. The three recently domesticated cage-birds from Australia, the zebra finch, budgerigar, and cockatiel, are still very common there. The ancestors of our major livestock birds—the jungle fowl (chicken), graylag goose (goose), wild turkey

Feral Pig Hunter in New Zealand *Pigs that have returned to life in the wild are of two kinds, the long-domesticated white hogs and the domesticated wild boars. They are established in many countries, where the white pigs soon grow long tusks and longer and darker bristles like their ancestral wild boars. In the United States and New Zealand, where the feral domestic pig above was shot in the Tararua Range, they are major game animals.*
Photo: Courtesy Department of Conservation, New Zealand. Crown Copyright.

(turkey), and mallard (duck), are also all very common animals, and the Crucian carp (goldfish), common carp (koi), and the Nile tilapia (tilapia) still thrive in their native waters. But the ancestors of several others are extinct or are now very rare. The progenitors of the horse (the tarpan), the cow (the aurochs), and the single-humped camel (the dromedary) are all extinct. The few surviving Nubian wild asses (*Equus a. africanus*), ancestors of the donkey, are seriously threatened by hunting and competition with livestock. The Cretan wild goat (*Capra aegagrus creticus*) ancestor of the domesticated goat, survives only in Crete's White Mountains and on two small offshore islands. Pure wild Bactrian camels are now very scarce, as their habitat has been infiltrated by feral animals. It is possible that the remaining wild Bactrian camels are feral animals anyway, and if so there is the added risk of dromedary blood being present as the two were often hybridized in central Asia. This still occurs in eastern Turkey, where both camels are kept, and hybrids are single-humped animals with heavy bodies and very thick legs. There is a similar risk to wild yak, as domestic yak and yak/cow hybrids are increasingly entering their limited remaining habitat on the Tibetan plateau.

There are many other examples of feral animals breeding with their wild conspecifics. Feral dogs are threatening the purity of the dingo in Australia. Donkeys breed with wild asses in the Horn of Africa, where feral ostriches have also joined flocks of wild birds. Except in the Kalahari Gemsbok Park, all the free ostriches in South Africa are now considered impure due to the escape of so many farmed birds, which were themselves hybrids. In northern Canada, reindeer introduced long ago to begin a Lapland-style herding industry have bred with the native caribou (both are subspecies of *Rangifer tarandus*). Feral cats have mated with wild bobcats in the United States and with their ancestor the African wild cat in southern Africa. Feral chickens in Southeast Asia have bred with their ancestor, the wild red junglefowl, and in many regions the junglefowl population is now impure.

Despite changes to their size and shape, color, and coat, most domesticated animals still resemble their wild ancestors. The many breeds of rabbits, for example, from the huge Flemish giant to the long-eared lop and the woolly coated Angora, are still unmistakably rabbits. Similarly, pigs are obviously pigs, from the white, long-bodied, bacon-producing Landrace to the blotched, drooping-eared Gloucestershire old spot. In fact, like several long-term domesticates, they are more recognizable to most people than their wild ancestors. When the opportunity arose to live wild and free again, these animals reverted to their ancestral appearance far more quickly than they attained their domestic characteristics. Although they evolved from purely domesticated stock, feral pigs in Australia and New Zealand now bear a close resemblance to their ancestor the wild boar, with plain or blotched dark coats, stiff bristles, and long tusks. In the United States they resemble wild boars even more closely, as they also have their blood from imports made in the nineteenth century for hunting. Feral rabbits in both New Zealand and Australia could be mistaken for those grazing alongside a hedgerow in an English field, and feral cats begin to resemble their wild ancestors after a few generations of freedom. Feral goats in the Canary Islands, believed to have been isolated there for 2,500 years, reverted to the ancestral uniform brown of the wild goat (*Capra aegagrus*) long ago. Feral chickens, now seen in many parts of the world, especially on islands such as Kauai, Bermuda, several Caribbean islands, and Guernsey, bear a distinct resemblance to their ancestor the red jungle fowl. However, the dingo, assumed to be a feral domesticated dog introduced into Australia between 3,000 and 4,000 years ago by the first colonists, bears no resemblance to the wolf.

■ PROBLEM ANIMALS

Generally, feral animals are problem animals; and the problem is manmade, caused by domestication in the first place. Some pet cats and dogs wander off, others are cast off. Other domesticates have escaped, but many were purposely released. Islands, large and small, have suffered most from these feral animals, which have wreaked havoc on their vegetation and native animals. Rabbits, sheep, goats, and pigs were liberated on remote islands centuries ago as a source of food for passing mariners, and eventually for the settlers. New Zealand, Australia, and Hawaii have the world's worse feral animal problems.

Rabbits accompanied the convicts to Australia in 1788, and were later taken to New Zealand. When they became a nuisance there, cats were released as biological controls, but they found native birds and small mammals more to their liking and easier to catch as they had no knowledge or experience of predators. In addition to the many aliens of nondomesticated origin, a host of former domesticated, and thus feral, animals now live freely in New Zealand. When the first European settlers arrived, the only native mammals there (two species of bats) had already been joined by black rats, feral goats and pigs introduced by sealers, whalers, and Captain Cook; and the Maori rat, which presumably accompanied the earlier Polynesian settlers. The herbivorous rabbits, sheep, goats, and fallow deer released by the settlers found the native tussock grass to their liking, while the kiwis, wekas, kakapos, and other ground-nesting birds were easy prey for the immigrant cats, dogs, and ferrets. The pigs rooted in the soil, destroying the native vegetation and causing erosion, and also ate bird's eggs and young. On sub-Antarctic MacQuarie Island, feral cats soon ate all the endemic red-fronted parakeets.

The vessels of the First Fleet and subsequent transportations, which carried convicts from England to Australia late in the eighteenth century, also carried other animals, several of which were to cause major problems for the island continent's plant and animal life. Rabbits, goats, and pigs were taken to the new colony for food, horses for transportation, and cats for company. All of them found the new land admirably suitable, and many became free settlers very quickly, some even before the convicts had served their terms. They were later aided in their colonization by the release or escape of others of their kind during the settlement of the new colony. The cats colonized many regions and helped to exterminate several small marsupials of the central deserts. Feral rabbits were the most serious pest, with millions devastating many regions prior to the introduction of the myxoma virus in 1950. Twenty million feral pigs, 2½ million goats and 300,000 horses also live wild and free. In addition there are believed to be five million donkeys roaming the more arid regions of the west, but the once massive water buffalo population of the seasonally wet regions of the north were almost eliminated by eradication campaigns in the 1980s. Camels were imported into Australia during the latter half of the nineteenth century, as beasts of burden in the dry outback, but with the advent of rail and road transportation in the 1930s, they were no longer needed and were abandoned. They thrived without human help in their new environment, which was very similar to their ancestral homelands. They can go without water for several days, and can store fat for even harder times. They eat the scrub vegetation, including species with a high salt content, and there is concern over their long-term impact on the vegetation, which evolved in the absence of large browsers. There are currently at least half a million feral camels in the outback, half of them in Western Australia, but they are all considered hybrids, as the original imports included single-humped dromedaries from India and Arabia and two-humped Bactrian camels from central Asia.

Hawaii has a similar history of the release or escape of domesticated animals that are established and threaten the unique vegetation and bird life, and Captain Cook is largely to blame. Pigs, sheep, and goats were off-loaded when his vessels *Resolution* and *Discovery* anchored in Kealakehua Bay on Hawaii's Kona coast in 1778. Captain Cook was later killed by angry natives when a storm forced him to return to the island. The sheep and goats were new colonists that soon became established, whereas the pigs (of European descent) joined the domesticated Asian pigs that the Polynesians had introduced, and eventually swamped their genes. Prior to the arrival of humans, Hawaii, like New Zealand, had just two native mammals, a bat and the monk seal. The climate, lush vegetation and lack of predators favored the new colonists, all of them extremely adaptable and very destructive. The vegetation on Laysan Island, in the northwest Hawaiian Islands, was completely devoured by introduced rabbits, causing the loss of several endemic birds. On other Hawaiian islands, pigs, goats, sheep, and cattle continue to ruin the native vegetation, affecting the survival chances of several native birds.

The new islands colonized by Europeans historically do not have the monopoly on feral animals, however. The fallow deer now freely browsing England's remaining ancient forests were taken across the English Channel by the Romans. Farmed arctic foxes in northern Scandinavia frequently escape, causing concern for the purity of the local wild foxes. Escapee mink from farms in Spain have been blamed for the disappearance of the Pyrenean Desman (*Galemys pyrenaicus*), a rare water shrew of fast-flowing mountain streams. Thousands of mink have also been released by activists, especially the Animal Liberation Front, and many are never caught. There were 69 recorded releases between 1995 and 2001 in the United States, and a fur farm in Washington suffered four releases in four years. Mink have also been released in Holland, Spain, Scotland, and England prior to the fur farming ban of 2000, and have established wild populations there, although they are rarely seen due to their nocturnal lifestyle. In Scotland there are currently estimated to be 50,000 feral American mink, and the ferret is also established there in the wild.

The domesticated animal with the longest history of freedom is the dingo, believed to be descended from dogs taken to Australia from Asia perhaps 3,500 years ago, according to fossil remains and aborigine rock drawings. It is thought that the early domesticated dogs in Eurasia may have resembled the dingo, but since then a great deal of selective breeding has produced modern breeds. It is therefore possible that the dingo was, until recently at least, the most purebred, domesticated dog in the world, its human-controlled breeding ceasing when it became feral. In recent years the increase of pet dogs that have more recently reverted to a feral life, and even domestic dogs from outlying homesteads, have diluted the dingo's genes, and pureblooded animals may only exist in the most remote regions and in the kennels of breeders who are preserving them. After so

American Mink   *The mink is one of the most successful colonists. Escapees and animals released from fur farms in England, Scotland, and several European countries are well established there, preying on native wildlife and endangering the water vole, Pyrenean desman, and the smaller European mink. In North America thousands of mink have been released from fur farms by activists.*
Photo: Photos.com

many years of freedom and natural selection, the dingo hardly deserves to be considered a feral animal anyway.

Although feral means living without direct human assistance, both cats and dogs have become parasites on humans in an urban setting. Feral dogs, usually strays that have banded together, have lived in short-term packs in many cities. Longer-term ferals, known as pariah dogs or pye dogs, are permanent fixtures in Indian cities and on their beaches, where they scavenge to survive. But dingoes are the only feral dogs capable of hunting for a living and able to find sufficient food. Similarly, feral cats are a problem in many cities in North America and around the world, living by scavenging, hunting rodents, or handouts. A feral cat colony with a difference is the one of homeless cats living on Parliament Hill in Ottawa, Ontario, Canada,where they are provided with shelters and food. But it is the rural feral cats that pose a major risk to native birds and mammals, especially in Australia, New Zealand, and the United States.

Feral animals do have some value, in addition to being the targets of sport hunters. They are considered important reservoirs of genetic material that may have been lost by their relatives still under man's control. The changes brought about by domestication result mainly from the selection of existing characters, not the production of new ones. When wild animals select their mates naturally in large populations, they deny the establishment of recessive genes, whereas pathological characters can be favored and selected in a small and closed captive population. In the reverse situation, when domesticated animals have become feral and are breeding naturally without the artificial selection practiced during human control, they contain gene pools of potential value to stock breeders.

■ SOME OF THE SPECIES

Mute Swan (*Cygnus olor*)

Mute swans are large whitewater birds with long necks. They stand about 4 feet (1.2 m) tall and weigh up to 33 pounds (15 kg), and are one of the largest flying birds. Although immediately recognizable as swans, they differ from the other North American species—the tundra swan and the trumpeter swan—by their orange bills, which have a black base and black knob over the bill. Although they were semi-controlled, pinioned, and harvested for food in England before the Norman Conquest (1066), their domestication really began in the twelfth century, when their popularity as a ceremonial dish increased and they were completely controlled. At the time all swans were considered the property of the Crown and became known as "royal swans." Eventually, the clergy and noblemen were also allowed to keep them, and so began a complicated system of marking them with notches on the webbing of their feet and on their bills to indicate ownership. Swans that escaped or avoided being marked were considered the property of the Crown. The marks were registered by a royal swan-herd, and the swans were rounded up at the annual "swan-upping." This practice of controlling them ceased in the late nineteenth century, but a symbolic swan-upping ceremony is still carried out annually on the river Thames by the Worshipful Companies of Vintners and Dyers, livery companies (trade associations with a very long history) of the City of London. The swans that were no longer cared for became feral, often frequenting the lakes in public parks, where they seek handouts in company with feral mallards, Muscovy ducks, and Canada geese. Mute swans are still favored display birds in zoos and bird gardens. The largest captive flock, at Abbotsbury Swannery in Dorset, England, contains over 1,000 swans and has been maintained there for 600 years.

Mute swans were taken to America in the late nineteenth century for zoos and for lakes on private estates and in public parks. Unpinioned swans escaped and became established in the wild, and were joined between 1910 and 1912 by several hundred swans that were released in the Lower Hudson Valley and on Long Island. These feral birds reached Connecticut and Pennsylvania in the 1930s, and in the 1950s were seen around the Great Lakes. They are currently well established along the Atlantic States Flyway from Maine to Florida, and have also reached the Pacific Northwest. They migrate locally, moving from frozen inland lakes to sheltered coastal bays for the winter. These feral mute swans are considered an invasive species, as they feed on submerged vegetation and soon overgraze the limited supply, especially in Chesapeake Bay where they have reduced the habitat for fish and crabs. They trample tern nests and are aggressive to the native tundra swans, driving them from their usual wintering grounds. They are extremely territorial and nasty when nesting and kill any bird approaching too close to their nest, even ducklings and goslings.

Pigeon (*Columba livia*)

The ancestor of the common domesticated pigeon was the rock pigeon (*Columba livia*). Reliefs show that pigeons were kept by humans in Persia and

Mesopotamia about 4000 BC, but it is unclear if they were wild birds or domesticated ones. The first indisputable evidence of large-scale breeding are the towers, called columbaria, that were built by the ancient Romans for pigeons to nest in, and from which their fat chicks or squabs were taken for eating. These towers date from early in the first century AD, and adult pigeons were apparently also stuffed with soaked grain to fatten them and had their wings clipped to prevent them from flying. Pigeons apparently also had religious importance at the time, and the Romans may have used them as carrier pigeons to send messages. Domesticated pigeons were introduced into North America by French colonists early in the seventeenth century. Modern-day pigeon keeping includes the homing pigeons that are used for long-distance races, and the many unusual breeds kept by pigeon fanciers in their lofts. Pigeons are still eaten. Commercial production is a small but well-established industry in California, where squabs are produced for the oriental market. In Europe a new commercial mutant pigeon has been produced called the euro-pigeon, with better conformity and improved meat quality.

The outcome of these many centuries of domestication was the inevitable escape of pigeons and the deliberate release of others. Their free-flying flocks have attracted and integrated expensive homing pigeons and rare breeds from pigeon fancier's lofts. Consequently, the feral pigeon is now a very common and cosmopolitan bird, attracted to city buildings for nesting and roosting, city squares and parks for

Feral Pigeons in Canada *The rock pigeon was possibly domesticated by the ancient Romans, and since that time escapees have become established in many countries, especially in urban areas, where tall buildings replace the natural rock canyons of their ancestors. They are considered pests, fouling buildings with their droppings and nests, and they are also accused of spreading disease.*
Photo: Clive Roots

handouts, and to rural grain elevators for spills. Some feral pigeons resemble their rock dove ancestors, but most are exceedingly variable in color, with a wide range of blue, gray-blue, reddish-gray, and mottled plumage. They interbreed with their ancestor the rock dove, and pure birds now only survive in remote regions that the feral pigeons do not reach. One of the most famous feral pigeon sites is London's Trafalgar Square, where their feeding is now prohibited; but on Mumbai's waterfront, in front of the famous Gateway of India, the forecourt is thick with grain put out for the numerous feral pigeons. They are known to carry several nasty zoonoses, such as psittacosis (ornithosis), encephalitis, aspergillosis, and coccidiosis, and anti-pigeon sentiment has been stirred in recent years by the claim that they may transmit the H5N1 strain of avian flu. In reality they are resistant to that virus, and from the general disease-transmission point of view they do not pose a significant health risk. But they are messy birds. Roosting and building their nests on the ledges of buildings, around rooftop air conditioning units, and under bridges, their droppings foul and deface buildings and make fire escapes and the sidewalks below slippery and unsafe. They block roof drains with their feathers and nest material, even though their nests are very simple affairs, and they attract rats and cockroaches.

Sheep (*Ovis aries*)

Sheep were one of the first livestock mammals to become domesticated, probably about 10,000 years ago adjoining the Fertile Crescent of the Middle East. The mouflon was the ancestor of all the modern breeds, of which there are more than 300, and the many surviving "primitive" breeds. In addition, purebred mouflon, which still resemble their wild counterparts, have also been kept and bred for many years and are thus considered domesticated. Sometime after their domestication long ago, sheep were introduced onto the Mediterranean islands of Corsica, Sardinia, Rhodes, and Cyprus, and the current populations there may be pure mouflon or the descendents of feral animals. Many of the wild mouflon in Europe may also have been infiltrated by feral sheep.

Sheep are very adaptable animals, able to survive wherever there is adequate vegetation, but especially in rocky regions. Escapees from hill farms, sheep abandoned when their farming was no longer profitable, and others deliberately released, have all established long-lasting feral populations, in many countries and on numerous islands. Feral sheep live on the island of Santa Cruz, off the coast of southern California, and on Ascension Island in the middle of the South Atlantic Ocean. Domesticated sheep were introduced into the Hawaiian islands by Captain Cook in 1778, and their feral descendents still graze the slopes of Hawaii Volcanoes National Park and can be hunted on Niihau. Domesticated mouflon were introduced at a later date and occur ferally on Hawaii and Lanai. There are also thousands of feral mouflon on Kahuku Ranch adjoining Hawaii Volcanoes National Park, where their ancestors were released in the 1960s for hunting. The ranch's 116,000 acres (47,000 ha) were purchased recently by the Nature Conservancy and the National Park Service—the largest conservation purchase in Hawaiian history.

Many primitive breeds of sheep exist on islands on which they were released over a century ago, and they are considered important depositories of genetic material, likely lost by modern breeds through selective breeding. Primitive breeds are those that have remained unchanged during the last two or three centuries of selection elsewhere, for prior to then breeding was mainly by natural selection. They have therefore retained many of their ancestral characteristics and are well adapted to their environment, which is usually a harsh one. Consequently they could be valuable for improving current modern breeds, such as improving their ability to live on marginal lands that are unsuitable for any other form of agriculture. In free-ranging populations (sheep, goats, or any other feral species), random mating and natural selection perpetuates the fittest genotypes. Reserves of animals, especially on isolated islands, where they have lived and bred without chance of contamination, is the best way to maintain a gene bank for future use. Inbreeding to "fix" desired characteristics in domesticated animals reduces genetic variability, the choice of genes being further reduced as selection for favorable characters continues. Eventually there may be a need for an infusion of new and unrelated blood from other genetic strains—especially of primitive or unimproved breeds, and possibly other species, in an attempt to produce hybrid vigor. In the words of IUCN's 1980 World Conservation Strategy, "the preservation of genetic diversity is both a matter of insurance and investment."

Several islands off the Scottish coast shelter the most primitive forms of sheep. Shetland sheep on the islands of the same name have evolved in virtual isolation since the eighth century. Soay sheep on the island of Soay in the St. Kilda Archipelago, are believed to have been there also from the same time period and were probably introduced by the Vikings. Orkney sheep are also believed to be descended from the Vikings' Scandinavian sheep, and were kept by the Scottish crofters before the depopulation of the highlands in the eighteenth century. Most of the surviving Orkney sheep live on the island of North Ronaldsay, where their diet is exclusively kelp thrown up onto the island's beaches by high tides. In 1974 The Rare Breeds Survival Trust bought the island of Linga Holm near Stronsay and released almost 200 pure Orkney sheep there from North Ronaldsay, where the reduction of the flock was necessary to avoid overcropping the seaweed. They do not have a uniform color and may be gray, white, or brown, and have been genetically isolated since early in the nineteenth century. Only natural selection has occurred since then, and they have been totally "wild" since the last islanders moved out in 1930.

Several primitive sheep breeds also survive in New Zealand. Mohaka sheep, of the Hawkes Bay region, have been uncontrolled since the early 1880s, and Hokonui sheep have been feral in the rugged hills of Southland since 1863. The ancestors of the feral sheep on Arapawa Island in Marlborough Sound, probably Australian merinos, were released there in 1867. Feral merinos also live on sub-Antarctic Chatham Island, descended from sheep taken there in 1841, but the large flock of several thousand feral sheep on neighboring Campbell Island, there since farming was abandoned in the 1930s, were all exterminated in the 1980s. The Australian outback, however, where sheep are actually farmed, is unsuitable for them to live ferally. Despite being overrun with feral herbivores and the enormous population of farmed sheep, there are no feral sheep there, due in part to predation by dingoes and feral dogs but also to their inability to survive there without help from man.

Water Buffalo (*Bubalus bubalis*)

The wild Asiatic buffalo is a native of India, Ceylon, and Burma, where it lives along the fringes of the jungle and in grassy clearings near water. In many places, it now breeds with the widespread and plentiful domestic animals, and those that have become feral and venture more frequently into the forest. The origins of man's control of the wild water buffalo are lost in antiquity, although the process may have started 3,000 years before the Christian era, as the animal is represented on seals used 5,000 years ago in what is now India. It was also recorded in China 1,000 years later. It reached Mesopotamia in about 2500 BC and many years later was brought from the Middle East to Europe by crusaders and pilgrims. While the domestic water buffalo has changed little from its wild ancestor in both appearance and habits, there is considerable variation in its size, the weight of bulls ranging from 660 pounds (300 kg) in China to over 2,200 pounds (1,000 kg) in Thailand, at least until recently.

There are two groups of genetically distinct water buffalo, the swamp buffalo (which has 48 chromosomes) and the river buffalo (with 50 chromosomes). Swamp buffalo are mainly animals of Southeast Asia, where they are called carabao; they are heavy, slate-gray, ox-like creatures with huge swept-back horns. They are the working beasts of the rice fields, used also for meat production but seldom for their milk. River buffalo live in southwest Asia, ranging from Burma through India and Mesopotamia into Europe. They are darker in color, often almost black, and usually have curled horns. River buffalo are quite cold-hardy, for they live in northern Afghanistan and Pakistan and on the cold plateau of Turkey. They have also been introduced into Europe, where they live as far north as 45°N latitude in Romania and almost as far north in southern Russia. In eastern Europe they are now common animals in several countries, including Bulgaria, Greece, Hungary, Yugoslavia, and Italy, where their milk is used to make quality mozzarella cheese.

Water buffalo are major beasts of burden and suppliers of meat and milk. They are farmed in East Africa, Madagascar, Papua New Guinea, Central America, South America, and Trinidad. They are also proving to be useful domesticated animals on farms in Florida and Louisiana, and have recently been imported into Canada. Their importation into Australia dates back to soon after the continent's European colonization. In the 1820s about 80 water buffalo, of both the river type and the swamp type, were imported into northern Australia. They were kept on Melville Island and on the mainland's Coburg Peninsula to provide food for remote settlements, but animals escaped; and when the settlements were abandoned in 1949, the uncontrolled buffalo spread across the northlands. Crocodiles ate the smaller animals, and the herds supported a large meat and hide industry, whose hunters culled between 5,000 and 10,000 animals annually. But they steadily increased and by 1985 numbered 350,000, with 20,000 living within Kakadu National Park, a World Heritage Site.

Water buffalo do well on poor forage, and studies have shown their rumen microorganisms digest cellulose (plant structural fiber) far more efficiently than cattle. They select a wider range of plants than cattle, eat sedges and rushes, and browse the leaves of water's edge trees. In northern Australia they have thinned out the formerly impenetrable thickets of pandanus palms by eating the shoots and juvenile plants; and they have actually been used to clear out woody plants, which cattle do not eat and which threatened to overwhelm their pastures. In the wet season the feral water buffalo's wallowing, trampling, and feces destroyed lagoons; they ate the floating lilies and water hyacinths and dove 6 feet (1.8 m) to graze on pond-bottom plants. In the dry season they ate the grasses that kangaroos and cattle depended on, and damaged trees by rubbing the bark. But the most unacceptable aspect of their presence was the risk they posed to the cattle industry from the brucellosis and tuberculosis they harbored. An eradication program was initiated in 1986, and all the buffalo in infected areas were slaughtered. The feral water buffalo remaining in disease-free areas number 20,000–30,000.

Goat (*Capra aegagrus hircus*)

Domesticated goats are the most environmentally damaging herbivores, tree-climbing browsers and bark-eaters that destroy vegetation very quickly. They are scrubland and alpine animals, and very hardy, able to survive in rough country during harsh winters. They are believed to have been domesticated from the wild goat (*Capra aegagrus*) about 10,000 years ago in Iran, at the edge of the Fertile Crescent, at about the same time that wild sheep were first controlled. Their domestication led to the production of many breeds, of which the angora goat is one of the oldest, developed in Turkey about 1500 BC. With selective breeding for wool quality and quantity, these goats can now produce 5 pounds (2.2 kg) of mohair twice annually, and they can be sheared like sheep, an unusual characteristic for a goat. The anglo-nubian is a more recently developed all-purpose goat, bred for milk and meat, while the Saanen, a Swiss dairy goat, is the highest milk-producing breed. Like sheep, some very ancient breeds of goats have survived, and have remained unselected and unchanged for centuries. The Bagot goat is one of these, living a semi-feral existence in a British Park, and able to select its mates naturally. Its ancestors were brought to Blithfield Hall in Staffordshire from the Rhone Valley by returning Crusaders in 1380.

Domesticated goats have become feral in many countries and on many islands, with devastating results for the native flora and fauna. They roam freely in Britain, Texas, the Canary Isles, the Galapagos Islands, Hawaii, Australia, and New Zealand. All the major Hawaiian islands except Lanai and Niihau have suffered from feral goats, descendents of those first introduced by Captain Cook in 1778. Even the remaining goats on Crete, usually considered a wild subspecies, may be feral descendants of domesticated goats taken there long ago. Britain's feral goats are believed to stem from escapees at least a century ago, and have not been improved by the infusion of modern breeds. About 2,500 feral goats live in the Scottish highlands, where they have coexisted with sheep for generations and are tolerated as they eat coarse vegetation such as sedges and ferns, which sheep reject, and therefore encourage more grass growth for the sheep.

Feral goats have been very destructive on islands in the eastern Pacific. The vegetation on Juan Fernandez Island off the coast of Chile, on which Alexander Selkirk (the real-life Robinson Crusoe) lived alone for four and a half years, has been seriously degraded by feral goats. In the Galapagos Islands, they denuded Santa Fe (Barrington) Island, whose tortoise was already extinct, but have since been eradicated and the vegetation is regenerating. On the island of Pinta (Abingdon), the endemic Abingdon tortoise (*Testudo elephantopus abingdoni*) was believed extinct, due to overkill by visiting mariners for food, and the last one was seen in 1906. Goats were released on the island in 1952 and multiplied, threatening the vegetation. An eradication program commenced in 1971 and within a decade 41,000 goats had been removed, and the remainder were killed between 1999 and 2003. During the campaign a single tortoise was discovered, the lone survivor of the subspecies, since named "Lonesome George." Across the world in the Indian Ocean, feral goats live on the two main islands of the Aldabra Atoll, home of the world's other giant tortoises. On the semiarid islands of Malabar and Grande Terre, which have no standing fresh water and infrequent rainfall, the goats must drink seawater most of the time.

Goats were first introduced into New Zealand by Captain Cook, but many were later released on marginal lands and on offshore islands, usually as a future source of food needing no care. They now occur throughout the country despite massive eradication campaigns that killed 332,000 between 1951 and 1958. Feral goats currently occupy almost 25 percent of the land controlled by the Department of Conservation, where their browsing destroys plants and causes erosion. The males have large curling horns and are a popular target for New Zealand's large alien animal sport hunting industry. Australia has a similar feral goat problem. They accompanied the convicts from London on the vessels of the First Fleet to provide milk and meat for the new colony. Visiting mariners later released goats as a future source of food. These animals, plus escapees from the colony, soon became established on the dry, rocky hillsides, and now occupy all six mainland states. Their population is currently estimated at about 2½ million. They damage the vegetation, kill smaller trees by bark-eating, and climb into higher trees to reach the leaves. They carry foot-rot and infect sheep, and then reinfect them after they have been treated, and they compete with native animals and farmed sheep for the sparse grasses. They are preyed upon by dingoes and feral dogs.

Donkey (*Equus asinus*)

The name donkey or burro (from the Spanish *borrico,* for donkey) applies to domesticated animals, while their wild ancestors are called asses. The donkey is descended from the African wild ass (*Equus asinus*), whose original range may have been the whole of North Africa, but this is unclear due to its virtual extermination and the presence of feral donkeys. The remaining small populations of these asses are restricted to the stony deserts of the most remote parts of southern Sudan, Ethiopia, and Somalia, but feral donkeys also live there and are known to interbreed with the asses. It is a very rocky and arid region, and the wild asses have long, thin legs and narrow hooves, suited for traversing the stony and hilly ground—unlike the semidesert-dwelling Asiatic asses, with their large and broad hooves for running on sand.

Wild asses are believed to have been first controlled by man about 4000 BC in Lower Egypt, and in the following millennium were used by the Sumerians of southern Mesopotamia to pull their chariots. They are mainly grazers, but like the feral donkeys studied in Arizona, which grazed and browsed in almost equal proportions, they no doubt take advantage of all available vegetation in their arid homeland. African wild asses have a shoulder height of 4 feet (1.25 m) and weigh up to 550 pounds (250 kg). Their coat is short, and it is reddish gray in summer, paler in winter, with white underparts all year. They have thin dark bands on their legs, an upright mane and tufted tail, and some have a dark dorsal stripe. Donkeys vary in size from the Sicilian miniature donkey to the large and shaggy-coated Poitu donkey. They usually have a dark stripe along the spine and a transverse one across the shoulders, but unlike the wild asses they lack the dark bars on their legs and show considerable variation in coat length. In addition to their presence in the Horn of Africa alongside their ancestors, feral donkeys also live on the island of Socotra, in the northwestern Indian Ocean, and on the mid-Atlantic island of Ascension. But their most well-known populations live in the American southwest and in Australia.

Donkeys entered the New World for the first time when Christopher Columbus arrived at Hispaniola in 1495 with the vessels of the Second Voyage (1494–96), which also carried horses, cattle, sheep, and goats. These donkeys were the founders of the line that produced animals for the conquistador's conquest of central America, and were later used as pack animals in the silver mines. During the establishment of the missions in the American southwest, donkeys were first seen in Arizona at the mission of San Xavier de Bac in 1679. During the mining boom in the Colorado River Valley (1858–80), they were used as pack animals by the prospectors, and as the mining camps closed, the burros were abandoned. Like their counterparts abandoned in arid Australia, they had no difficulty settling in the arid lands, which resembled their ancestral rangelands. Prior to 1969 thousands of burros were culled. They became a nuisance in Death Valley National Monument, where they numbered 2,000 in the early 1980s, and fouled the water holes and competed with the endangered desert bighorn for the meager vegetation. Excess burros are currently rounded up and offered to the public through the Adopt-A-Burro program.

Donkeys were also used as pack animals in Australia's dry outback beginning in 1866, especially in regions where certain plants were toxic to horses. They were abandoned to their fate when motorized transportation arrived, but they were more than equal to the hardships of desert survival and thrived in the outback. In 2005 Australia's feral donkey population was estimated at almost five million, mostly in the arid center and the northwest.

Horse (*Equus caballus*)

The only truly wild horse is the Przewalski horse, now surviving in zoos, and in the Gobi Desert where captive-born horses have been reintroduced into their former habitat. All the other horses that run wild, on islands such as Assateague, Sable, and Fraser, in the American southwest, and in New Zealand, are all feral animals, the descendents of domesticated horses that were released or escaped. Feral horses

Feral Horses (Mustangs) in the United States   *Domesticated horses have been released, or escaped from their paddocks, and are now established in several countries and on a number of islands. The mustangs of the American southwest are the most well known of these, descendants of horses that escaped from the Spaniards, early settlers, and the U.S. cavalry. Symbols of the freedom of the west, they now number about 25,000.*
Photo: Robert Broadhead, Shutterstock.com

have existed in New Zealand for 150 years. There are currently two populations, one of at least 1,000 horses in the Kaimanawa grasslands of the Army Training Area south of Mt. Ngaruhoe in North Island, and a smaller population in the Aupouri Forest in the extreme northern part of the island. The wild horses of Sable Island, a wind-swept, grass-covered sandbar in the Atlantic, 185 miles (300 km) southeast of Nova Scotia, have a much longer history. Thomas Hancock, a Boston ship owner, liberated 60 horses on the 26-mile-long (42 km) island in 1760, and others were released in the 1800s and early in the twentieth century. The island's horse population fluctuates between 200 and 350 animals.

The most famous wild horses are the mustangs of the American southwest, whose name is derived from the Spanish *musteño,* meaning wild. They were originally feral Spanish horses, but their blood has been diluted by other breeds and much mingling since their early days. Some of the new blood came from escaped cavalry horses, which the American government had improved with purchases of German East Friesian stallions. French horses, escapees from settlers in the Detroit and New Orleans regions, were pushed westwards by colonization and mingled with the western herds, eventually all being driven into the more arid lands of the west. There are currently about 25,000 mustangs, mainly in Nevada, Montana, Colorado, and Utah; there are also a few in Alberta and British Columbia.

Australia's wild horses are called brumbies, after an Englishman who served in the army guarding convicts in New South Wales in the late eighteenth century. Like other soldiers there, he was given a tract of land on which he kept a few horses, but when he was posted to Tasmania in 1804, the horses were abandoned to their fate. They survived and aided by other escapees and liberated animals gave rise to the feral herds that are mainly concentrated in Western Australia, the Northern Territory, and Queensland, with a few in the Snowy Mountains.

Rabbit (*Oryctolagus cuniculus*)

The natural range of the rabbit is the Iberian Peninsula, where it lives in dry, grassy areas with sandy soil suitable for burrowing. It was introduced to the rest of Europe and some Mediterranean Islands by the ancient Romans, and later to many other parts of the world. A grayish-brown animal, known for its highly reproductive capacity, the rabbit can breed year-round, with litters averaging six naked and helpless young in an underground nest, after a gestation period of 30 days. The Romans are credited with first breeding and therefore domesticating wild rabbits, probably in the first century BC, when they kept them as a source of food in walled courtyards called leporia. They may have practiced selective breeding for color and size, but by 600 AD rabbits were also being raised as food in British monastery gardens by the monks, who apparently preferred newborn young and even well-developed fetuses. The first selective practices for size, tameness, and perhaps color possibly occurred then. In the nineteenth century, rabbit fanciers modified breeds for wool or meat, and then toward the end of the century, in the Victorian Era, they became popular as pet and show animals. Many breeds are now recognized, including the largest meat-producing one—the giant chinchilla—which may weigh 16 pounds (7.2 kg); the rex, which is bred for its velvet fur; and the angora, for its long wool. Hares have never been domesticated; the Belgian hare is a breed of rabbit.

During the age of discovery and the following age of settlement, domesticated rabbits were carried in ships as food and were released on many islands as a future source of food for visiting mariners. Their ability to revert to a feral life and to their normal, wild, coloration, has been well demonstrated by these animals, for they became established in many parts of the world and proved to be one of the most destructive of all feral animals. Australia and New Zealand have suffered most from the rabbit's rapid multiplication and spread. Like goats, sheep, and horses, rabbits were also brought to Australia by the First Fleet carrying the convicts from London, and within a few years escapees were well established around Port Jackson (later to be renamed Sydney). They were also liberated in Tasmania and were abundant there by 1827. Other releases of rabbits on the mainland resulted in their spread across the southeast corner of the continent by 1859, and by early in the twentieth century they had colonized the whole country in suitable habitat (excluding the extreme north) and numbered an estimated 20 million. They crossed the desert and reached Western Australia, despite the erection of a 1,000-mile-long (1,650 km) rabbit-proof fence to keep them out. Cats were first used as biological controls to reduce the rabbits, but they could not cope with the large numbers and eventually themselves became serious pests. Myxomatosis, caused by the rabbit-specific myxoma virus, was introduced in 1950, and killed 90 percent of the feral rabbits in the drier parts of the country. After extensive trials, and its escape from a quarantine facility, rabbit calicivirus, also known as Rabbit Hemorrhagic Disease, was approved for release in 1996. It proved very effective in killing rabbits in the moister regions of the country. There are still feral rabbits in Australia, but they are no longer the serious pests of just a few years ago. They are killed by foxes, reducing their predation on small marsupials, and the aborigines also hunt them for food. In New Zealand, feral rabbits are still a serious problem in South Island, where they compete with the sheep for the tussock grass, cause soil erosion, and aid the spread of noxious plants. Elsewhere they are held in check by stoats, feral cats, and ferrets.

Pig (*Sus scrofa*)

There are two types of domesticated pigs, European breeds that are descended from the European wild boar (*Sus s. scrofa*), and Asiatic or oriental breeds that originate from another subspecies, *Sus scrofa vittatus,* that lives in the Far East. Their domestication goes back to at least 7000 BC, separately in the East and the West. The arrival of domesticated pigs in Australia dates from the time of the first European settlement, and by the late nineteenth century escapees were established in several regions and then spread rapidly over the whole continent, excluding the dry center and west. The latest estimate of their numbers is 20 million; they damage lagoons and floodplains with their wallowing and rooting, and have been accused of spreading leptospirosis and foot-and-mouth disease. Pigs were first introduced into New Zealand by the colonizing Polynesians, then by the French explorer De Surville in 1769, and Captain Cook in 1773, as gifts to the Maoris. Their feral free-living descendants are still known as "Captain Cookers." Their descendents flourished and at one time numbered 120 per square kilometer in suitable habitat. They now have thick, bristly, dark-brown or blotched coats, but their piglets have grayish coats and are usually stripeless, unlike those of their ancestors. They have been the object of intense control campaigns and are a major sport hunting animal.

Domesticated pigs were first introduced into the contiguous United States by the Spanish explorers, and then by settlers, as early as 1539 in Florida; and escapees, plus released and abandoned animals, gave rise to the large feral population. But many differ from the feral pigs in Australia and New Zealand because wild boars were also introduced into New England, Arkansas, and Tennessee in the nineteenth century for hunting, and the feral pig population therefore has a mix of wild boar and feral domestic pig blood. They bear a greater resemblance to wild boars, with dark coats, a heavy neck, a shoulder mane of bristles, and upper tusks that may reach 9 inches (23 cm) in length; their piglets are often striped like those of the wild boar. Now established in 20 states, feral pigs are the most common feral ungulates in the United States, with a population estimated at 4 million and large numbers in Texas, Florida, and California. They may be called wild hogs, or "razorbacks," and are very destructive animals, for they eat amphibians, reptiles, small mammals and nesting birds, and all manner of vegetation. They destroy shallow pond ecosystems with their wallowing and defecating, and they carry leptospirosis, brucellosis, and bovine tuberculosis, making them a potential hazard to the livestock industry. Feral pigs occupy all the Hawaiian islands except Lanai and Kahoolawe, where they have been exterminated. The original stock on the islands was of Asian descent, from the domesticated pigs brought by the first Polynesian settlers. Captain Cook then liberated domesticated European pigs in 1778, and others were released later by settlers, and these diluted the Asian blood. Their rooting has destroyed the native ground vegetation, prevented its regeneration, and opened up areas to invasion by alien plants.

In addition to the feral domesticated hogs, and hybrids between them and wild boars, there now also feral populations of pure wild boars (*Sus scrofa*). They have been farmed in several countries for their lean meat, but were kept pure and have retained the characteristics of their wild ancestors. These animals are beginning to escape from their enclosures, just as the domesticated hogs have done for years. Wild boar introduced as an alternative farm animal in Saskatchewan in the 1980s, escaped recently and are established in Moose Mountain Provincial Park, where the extreme winter climate is similar to that of their ancestor's Siberian homeland. Escapees are also established in England, the first wild boars to live freely there for three centuries.

Ferret (*Mustela putorius furo*)

The ferret is a domesticated European polecat (*Mustela putorius*), or possibly a steppe polecat (*M. eversmanni*), and is therefore a close relative of the mink and weasels. Its domestic origins are lost in antiquity, but its initial value to man was apparently as a control of vermin. Ferret's bones found near human settlements have been dated to 1500 BC, but how they were differentiated from those of the wild polecat is unclear. They were apparently used during the reign of Emperor Caesar Augustus (64 BC–14 AD), to control a plague of rabbits on the Balearic Islands. Ferrets were kept in Spain at the time of the Moorish conquest in the eighth century AD and spread throughout the Mediterranean regions. From the Middle East they were brought back to England by returning crusaders, to control rats in their manor houses. Beginning in the fourteenth century they were used by English poachers to flush rabbits out of their burrows, during which time albinos (actually yellowish-white animals) were selectively bred so they could be seen at night. During my childhood in the Welsh hills I used them for rabbiting. An albino ferret is depicted, but misidentified, in Leonardo da Vinci's 1485 painting titled *Lady in Ermine*. Queen Elizabeth I had a pet ferret. In recent years ferrets have been used as laboratory animals and have become increasingly popular as house pets; one is the regimental mascot of Britain's 1st battalion, The Prince of Wales's Own Regiment. Ferrets are farmed for their pelts in Finland.

Like their ancestors, ferrets are highly carnivorous and need a diet of animal protein, but despite claims to the contrary, they can survive in the wild if food is available, as many escaped or abandoned former pets have proved. Feral ferrets, about 2,500 strong, live in Scotland, and a population of ferrets survived on Washington's San Juan Island in the Strait of Juan de Fuca for many years. They have also survived in southern Alaska, but it is in New Zealand that they have shown how easily they may become reestablished in the wild, even after many centuries of domestication.

Ferrets were deliberately released in New Zealand in the late nineteenth century to control the introduced rabbits. A few were imported in 1879, followed by thirty two shipments from London in 1882–83, totalling 1,200 animals. Only half of

Feral Ferret in New Zealand *Several thousand ferrets were deliberately introduced into New Zealand, beginning in 1879, to control the previously introduced rabbits, which had reached pest proportions. They flourished and, despite intensive trapping campaigns, have severely affected the native birds, especially ground nesters such as the kakapo and penguins.*
Photo: Courtesy Department of Conservation, New Zealand. Crown Copyright. Photographer: Rod Morris, 1982

these animals arrived alive. These were said to have been mated on the voyage with pure, wild European polecats, creating hybrids that would have greater survivability in the wild. However, capturing large numbers of polecats in the first place would have been very difficult, and breeding them to ferrets during a long sea voyage, with obviously very primitive caging, would have been well nigh impossible. It seems more likely that the ferrets of the time, as they are today, were of the two color phases—the albino and the polecat—which resembles its wild ancestor's color, and feral animals generally revert back to this ancestral coloration. Between 1884 and 1886, almost 4,000 ferrets were released, despite concerns that they would prey on native wildlife when the rabbits were reduced, which is exactly what happened. The country's ground-dwelling birds have suffered severely from the feral ferrets, together with the wild stoats and weasels that were also released. The have killed many of the very rare kakapos and yellow-eyed penguins, and as 30 percent of the ferrets carry bovine tuberculosis, they are a threat to the cattle industry.

Cat (*Felis catus*)

Wherever cats are kept, they have become feral. Around the world they have shown their typical independence and returned to a successful life in the wild. They may begin simply through being abandoned, but they may also just go off on their own, find prey plentiful, and never return home. In the United States there is a large population of feral cats, but uncertainty over what action, if any, should be taken to control them. They are accused of killing millions of native birds annually, yet they also keep rodents in check. They are in turn killed by foxes and coyotes and can legally be shot in some states, but not in others. In other countries, especially

Australia and New Zealand, there is no ambiguity about feral cats; they are considered pests of the worst kind and must be controlled.

House cats are descendents of the African wild cat (*Felis libyca*), but they can interbreed with several other species, and where these animals still occur in the wild, they are at risk of hybridizing with feral cats. In Scotland, after interbreeding with feral domestic cats for two millenia, it is thought that at least 80 percent of all the European wild cats are now hybrids. Others believe that there are no pure wild cats left. In South America feral house cats have hybridized with wild Geoffroy's cats, and in the United States they have mated with wild bobcats. The kittens of feral cats can be resocialized if they are hand-raised from an early age, but it is very rare for an established feral cat to return to the uninteresting and controlled life of a house cat.

The effect of feral cats on native wildlife is a major concern in the United States. Ground-nesting birds, such as plovers, quail, and terns, and even birds that nest in low bushes, are totally helpless. Most species of small mammals, from squirrels to mice, and especially nocturnal ones, have no defense against cats. The descendents of cats that were released or escaped from vessels visiting the Hawaiian Islands long ago, which have since been augmented by stray house cats, have had a devastating effect on the island's endemic birds. Colonies of ground-nesting shearwaters and petrels have either been totally destroyed by feral cats, or have been unable to raise any chicks due to cat depredations.

The cat's history of settlement in Australia is believed to stem from shipwrecked Dutch vessels in the seventeenth century, for they were known to the aborigines when the first European settlers arrived, with their own cats. Like cats everywhere these colonists could not resist the local wildlife, and by the mid-nineteenth century they were established as totally self-reliant feral animals, even in the "bush" miles from human habitation, on the mainland and in Tasmania. They were aided in their spread by the multitude of rabbits, but their large numbers are now more dependent upon native small mammals and birds. In addition to their very serious direct impact on wildlife, they transmit infections such as toxoplasmosis and sarcosporidiosis, a parasitic infection of the muscle that causes significant losses of livestock and native mammals, and for which cats act as the intermediate host.

Glossary

Aberration
An animal that deviates in important characteristics from its closest relatives.

Allele
Any of two or more alternate forms of a gene that occupy the same position on a chromosome. Inherited from each parent, they control a particular trait.

Anadromous
Fish that hatch in freshwater, migrate to the sea to grow, then move back to freshwater, usually their ancestral river, to breed.

Aquaculture
The controlled cultivation of aquatic plants and animals, including algae, shellfish, and fish—both freshwater and marine—when it is called fish farming

Artificial insemination
The technique whereby semen is collected from males and introduced artificially into the female's reproductive tract.

Artificial selection
The human selection of genetic traits, when animals are intentionally chosen for breeding to achieve or eliminate a specific trait. The animals' hidden recessive genes are then exploited, especially by inbreeding, producing variations dependant on domestication and providing the basis for change from the wild ancestor.
Synonymous with Selective Breeding. Antonym of Natural Selection.

Biomedical
Relating to the activities and applications of science to clinical medicine.

Breed
An artificially produced population of animals, developed from a domesticated species, through the selection of uniform characteristics of value to man. Breeds are the equivalent of the subspecies of wild animals.

Carnivore
Members of the order *Carnivora,* which include the cats, dogs, and hyenas. However, it is also used to denote any animal that eats animal protein, such as fish, crustaceans, and the meat of amphibians, reptiles, and mammals. When members of the same species are eaten, it is considered cannibalism.

Cecum
The cecum is a blind intestinal sac, the first portion of the large intestine. It contains micro-organisms that ferment complex carbohydrates, producing fatty acids which are then absorbed through the cecum wall. Captive grouse receiving inadequate roughage to replace their heather-and-pine-needle natural diet experienced shortening of the cecum by one quarter within one generation.

Chromosomes
Strands of genes contained in the nucleus of a cell, normally appearing in corresponding pairs, that determine growth and appearance.

CITES
The Convention on International Trade in Endangered Species. An international agreement banning commercial trade in many endangered species and monitoring and regulating trade in others that are at risk.

Cloaca
The organ into which an animal's digestive, urinary, and reproductive systems empty, opening via the anus.

Conspecifics
Individuals of the same species

Diploid
A full set of genetic material consisting of paired chromosomes, one from each parental set. Most animal cells, except the gametes, have a diploid set of chromosomes.

DNA
Deoxyribonucleic acid, the material inside the nucleus of cells that carries the genetic information.

Domestication
The continual control and breeding of wild animals for man's benefit, eventually resulting in changes to their genetic makeup and appearance.

Domesticates
Animals that have been or are being altered to suit man's needs through the process of domestication.

Dominant gene
A gene that determines the phenotype or physical characteristics. When a genetic trait is dominant, an animal needs to inherit only one copy of the gene for the trait to be expressed. Dominant genes have a 50 percent chance of passing from parent to offspring.

Endocrine glands
Glands that produce and secrete hormones into the bloodstream to regulate the body's normal functions.

Estivation
Long-term summer sleep to avoid hot and dry weather and to conserve water and energy, while surviving upon body fat.

Feral
Former domesticated animals that have become reestablished in the wild.

Gametes
Male or female reproductive or germ cells—the sperm and the egg. They possess the haploid number of chromosomes (half the number of the somatic or body cells).

Gene
The gene is the basic unit of inheritance, controlling the structure and function of organisms. It contains the discrete hereditary units that duplicate each time a cell divides, producing exact copies that determine the individual's characteristics. The selection of one allele over the other occurs purely by chance, so problems can occur in small populations when breeding involves only a few animals. Genes are transmitted to the next generation in the gametes or egg cells, and every baby animal receives a set of genes from each parent, which guide its development and enable it as an adult to live in the manner of its ancestors.

Gene pool
The collection of genes available among the breeding members of a population. In a captive closed population the gene pool is totally dependant upon the genes of the original wild animals—the founders.

Genetic diversity
Genetic variation within a population or species that results from differences in their hereditary material. It is essential for populations to be able to adapt to changing conditions, and for individuals to maintain their fitness.

Genetic drift
The loss of gene diversity in a population, drift is a random process in which some genes are not passed from parent to offspring and are then lost. In the wild, these genes can be recovered only through breeding with other members of the population (which is impossible in cases of population fragmentation), or through the very slow process of mutating.

Genetics
The scientific study of heredity.

Genome
The total chromosome complement of an organism.

Genotype
The genetic identity of an individual.

Genus
A biological classification; a group of living organisms having one or more related or morphologically similar species. The plural is genera.

Gonadotropins
Pituitary hormones that stimulate the reproductive system. They have been used artificially to stimulate egg-laying in farmed fish, and to encourage aquatic axolotls to become terrestrial.

Haploid
A single set of chromosomes (half the full set) present in the egg and sperm cells of animals.

Heredity
The transfer of characteristics from parents to offspring through their genes.

Heterozygous
When an animal's two copies of a gene have different alleles and it therefore has genetic diversity. Heterozygosity is necessary to maintain vigor and a viable population, and allows animals to adapt to evolutionary and environmental change. A genetically diverse population will have some heterozygous and some homozygous animals.

Heterozygote
An animal with two different alleles of the same gene, one normal and one mutant.

Homozygous
When an animal has two identical alleles for a particular mutation, so it lacks genetic diversity. The alleles can be both dominant or both recessive.

Homozygote
An organism derived from the union of genetically identical gametes and having identical alleles for one or more genes.

Hybridize
To breed with a member of a different species or subspecies. The offspring of subspecies are fertile, but in the case of hybridizing species fertility depends upon the closeness of the relationship of the parents.

Hybrid vigor
The increased size or fitness of a hybrid over that of a purebred animal, especially in fertility and survivability.

Inbreeding
The mating of biological first-degree relatives whose ancestors are shared. Inbreeding results in a decline of both population and individual genetic variation and is therefore harmful. Its detrimental effects result from the loss of beneficial genes combined with the expression of deleterious genes, which would have been swamped in a larger population.

Inbreeding coefficient
The degree of inbreeding, varying from 0 percent to 100 percent. It indicates the degree of probability that both alleles for any gene are the same, by descent, and the increase in homozygosity.

Linebreeding
The mating of individuals of a particular family line, such as first cousins, or uncle and niece, without causing high levels of inbreeding (so not first-degree relatives), in order to maintain the high contribution of a particular ancestor through successive generations.

Metabolism
The chemical processes within the body that sustain life. Some substances are broken down from ingested foods to provide energy, others are synthesized internally by the animal. Two main processes are involved, anabolism, which is the biosynthesis of complex organic substances from simpler ones, and catabolism—the breakdown of complex substances to release their energy.

Metabolic rate
The rate at which an organism transforms food into energy and body tissue. The amount of energy liberated per unit of time.

Mitochondria
Structures in a cell's cytoplasm that contain enzymes for cell metabolism, especially for converting food into energy.

Mouth brooder
Fish that hold their eggs, and then the fry or hatchlings, in their mouths for their protection. As they are cold-blooded, they cannot provide warmth.

Mutation
A change in a cell's DNA sequence, caused by copying errors in the genetic material during cell division, which can be inherited if it occurs in cells that produce sperm or eggs. It results in a mutant animal that differs from its parents in some manner, although this may not always be obvious.

Mutant
An animal that has been genetically changed as a result of mutation and differs from the normal or "wild type" of its species. It is a source of genetic variability, and naturally occurring genetic mutants are an integral aspect of evolution.
Synonymous with morph; also with mutation.

Mutant gene
There are thousands of genes in each parental set, which pass unchanged from generation to generation when functioning normally, but occasionally a gene is defective and cannot perform its normal function. Known as a mutant gene, like normal genes it is transmitted from parent to young in its defective state, producing variations to the normal genotype and phenotype. As every embryo carries genes from both parents, mating a mutant animal back to its parent will produce 25 percent mutants, 50 percent normal but carrying the mutant gene, and 25 percent purebred young.

Morph
A distinct genetic population of a species that is distinguished from others of its conspecifics by its color or pattern, but can interbreed with them.
Synonymous with mutant.

Morphology
The form and structure of an organism.

Natural Selection
When animals can select their own mates.
Antonym of Artificial Selection.

Nucleus
The part of a cell containing DNA (deoxyribonucleic acid) and RNA (ribonucleic acid), which are responsible for growth and reproduction.

Phenotype
The observable characteristics of an animal—its color, pattern, coat length, eye color, etc.

Physiology
The branch of biology dealing with the functioning of organisms.

Pituitary
An endocrine gland situated at the base of the brain that secretes important hormones, including the one responsible for growth.

Population bottleneck
An event in which a large percentage of an animal population is killed or prevented from breeding. Bottlenecks increase genetic drift and thus inbreeding due to the reduced number of potential mates. When a few animals are separated from the main population, this genetic bottleneck is called a founder event.

Precocial
Young animals that are well developed and mobile, and that have their eyes open and a body covering of down feathers or fur, at birth or hatching. They may be independent, semi-independent, or totally dependent upon their parents.

Race—see Subspecies

Random breeding
The absence of any reproductive system, when animals are chosen for breeding without regard for their parents or earlier ancestry.

Recessive gene
A gene that does not express itself when paired with a dominant gene. It is phenotypically manifest in the homozygous state, but is masked in the heterozygote by the dominant gene.

Selective breeding—see Artificial Selection

Sexual Dimorphism
Obvious external differences between the male and female of the same species.

Species
A biological classification of a naturally interbreeding population. A subspecies or race is a geographically separate subunit or segment of a species. When classifying domesticated animals breeds are the equivalent of the subspecies of wild animals.

SSPs
Species Survival Plans, begun in 1981 by the American Association of Zoos and Aquariums (AZA) to help ensure the survival of selected wild species.

Subspecies—see Species

Trait
A genetically inherited phenotypic (observable) feature of an organism, such as a thicker coat or longer horns.

Velvet
The soft covering of antlers during their growth stage, when they are richly supplied with blood vessels, and can be easily damaged. Antlers are removed while "in velvet" for medicinal purposes.

Bibliography

Ajayi, S.S. Wildlife as a source of protein in Nigeria: some priorities for development. *The Nigerian Field* 36, no. 3 (1971): 115–127.

Amphibiaweb. http://amphibiaweb.org/declines/zoo/index.html.

Barnett, S.A, and Stoddard, R.C. Effects of breeding in captivity on conflict among wild rats. *Journal of Mammalogy* 50 (1969): 321–325.

Beebe, Frank L. *The Compleat Falconer*. Hancock House Publishers, Surrey, BC. 1992.

Berry, R.J. Genetics and Conservation. In *Conservation in Perspective*. Edited by A. Warren and F.B. Goldsmith. Wiley, Chichester, 1983.

Bouman, Jan. The future of the Przewalski horses *Equus przewalskii* in captivity. *International Zoo Yearbook* 17, Zoological Society of London, London, 1977.

Breeds of Livestock. http://www.ansi.okstate.edu/breeds/

Bristol Aquarists Society. http://www.bristol-aquarists.org.uk.

Cultured Aquatic Species Information Programme. http://www.fao.org.figis/servlet/static?dom=root&xml=aquaculture/cultured_search.xml

Clutton-Brock, Juliet. *Domestic Animals from Early Times*. University of Texas Press, Austin, 1981.

———. *A Natural History of Domesticated Animals*. Cambridge University Press, 1999.

Darwin Crocodile Farm. http://www.crocfarm.com.au/australia.asp.

Delacour, Jean. *The Pheasants of the World*. Spur Publications, Hindhead, 1977.

Dharmananda, S. Endangered Species Issues affecting turtles and tortoises used in Chinese medicines. http://www.itmonline.org/arts/turtles.htm.

FAO Fishery Country Profiles. http://www.fao.org/fi/fcp/fcp.asp

Farm raised channel catfish. http://edis.ifas.ufl.edu/pdffiles/FA/FAO1000.pdf

Feral Animals in Australia. http://www.deh.gov.au/biodiversity/invasive/ferals/

Flores-Nava, A. *Cultivated Aquatic Species Information Programme - Rana catesbiana*. Inland Water Resources and Aquaculture Service. Fisheries Global Information System, 2005

Fox, M.W. *The Dog. Its domestication and behavior*. Garland STPM Press, New York & London, 1978.

Global Amphibian Assessment. http://www.globalamphibians.org/

Glofish. http://www.glofish.com

Greig, J.C., and Cooper, A.B. The "Wild" Sheep of Britain. *Oryx* X (1970): 383–388.

Guiness World Records. http://www.guinessworldrecords.com/

Hediger, H. *Wild Animals in Captivity*. Butterworth, London, 1950

Hoffmann, R.R., and Matern, B. Changes in gastrointestinal morphology related to nutrition in giraffes *Giraffa camelopardalis*. *International Zoo Yearbook* 27, Zoological Society of London, London, 1988.

Hsieh-Yi, Yi Chiao, Yu Fu, Maas, B., and Rissi, Mark . *Dying for Fur. A Report on the Fur industry in China*. 2006. http://www.animal-protection.net/furtrade/report_fur_china.pdf

King, J.M., and Heath, R. Game domestication for animal production in Africa. World Animal Review 16 (1975): 23–30

Jarman, M.R., and Wilkinson, P.F. Criteria for animal domestication. In *Papers in Economic Prehistory*, ed. E.S. Higgs. Cambridge University Press, 1972.

Lorenz, K. *Evolution and the modification of behaviour*. University of Chicago Press, Chicago, 1968.

Luxmore, R. Game farming in South Africa as a force in Conservation. *Oryx* XIX (1985): 225–231.

Monodelphis domesticus. The Laboratory Opossum. http://www.sfbr.org/pages/genetics_projects.php?p=51

Moss, R. Effects of captivity on gut length in red grouse. *Journal of Wildlife Management* 36 (1972): 99–104.

New Zealand's Rare Breeds http://www.rarebreeds.co.nz/

Policy Statement on Captive Breeding, International Union for the Conservation of Nature, Gland Switzerland, 1987.

Ratcliffe, Herbert L. Diets for Zoological Gardens: aids to conservation and disease control. *International Zoo Yearbook* VI, Zoological Society of London, London, 1966.

Ralls, Katherine, Brugger, Kristin, and Glick, Adam. Deleterious effects of inbreeding in a herd of captive dorcas gazelle *Gazella dorcas*. *International Zoo Yearbook* 20, Zoological Society of London, London, 1980.

Roots, Clive. Conservation by Domestication. *New Scientist* 47 (1970): 638–640.

Shi, Haitao, and Parham, James Ford. Preliminary Observations on a Large Turtle Farm in Hainan province, Peoples Republic of China. *Turtle and Tortoise Newsletter* 3 (2001): 4–6.

Sinha, V.R.P. Integrated Carp Farming In Asian Countries. 1985. http://www.fao.org/docrep/field/003/AC2363E/AC236#00.htm.

Slatis, H.M. An analysis of inbreeding in the European Bison. *Genetics* 45 (1960): 275–287.

Stanley Price, M. Fringe-eared Oryx on a Kenya Ranch. *Oryx* XIV (1978): 370–373.

The Tortoise Trust Library. Http://www.tortoisetrust.org/articles/articles.html.

Treus, V., and Lobanov, N.V. Acclimatization and domestication of the eland *Taurotragus oryx*. *International Zoo Yearbook* 11, Zoological Society of London, London, 1971.

The Welfare of Animals kept for Fur Production. http://www.bontvoordieren.nl/english.

The World Conservation Strategy. International Union for the Conservation of Nature, Gland, Switzerland, 1980.

Thorogood, June, and Whimster, I.W. The Maintenance and breeding of the leopard gecko *Eublepharis macularius*. *International Zoo Yearbook* 19, Zoological Society of London, London, 1979.

Tilapia Farming. http://www.ag.arizona.edu/azaqua/ata.html.

Watanabe, M.E. The Chinese Alligator: is farming the last hope. *Oryx* XVII (1983): 176–181.

Whitaker, R. Captive Breeding of Crocodilians in India. *Acta Zoologica et Pathologica Antverpiensia* 78 (1984): 309–318.

Wilkinson, P.F. Current experimental domestication and its relevance to pre-history. In *Papers in Economic Prehistory*, ed. E.S. Higgs. Cambridge University Press, 1972.

Wood, F. Turtle Culture. http://www.turtle.ky/scientific/culture.htm#fre.

Zeuner, F.E. *A History of Domesticated Animals*. Harper and Row, New York, 1963

Index

About the Author

A zoo director for many years, CLIVE ROOTS has made live animal collecting expeditions for zoo conservation programs to several regions, including Amazonia and remote islands in the South Pacific. He has acted as a masterplanning and design consultant for numerous zoological garden and related projects around the world, and has written many books on zoo and natural history subjects.